EVALUATION OF THE NATIONAL AEROSPACE INITIATIVE

Committee on the National Aerospace Initiative

Air Force Science and Technology Board

Division on Engineering and Physical Sciences

NATIONAL RESEARCH COUNCIL
OF THE NATIONAL ACADEMIES

THE NATIONAL ACADEMIES PRESS
Washington, D.C.
www.nap.edu

THE NATIONAL ACADEMIES PRESS • 500 Fifth Street, N.W. • Washington, DC 20001

NOTICE: The project that is the subject of this report was approved by the Governing Board of the National Research Council, whose members are drawn from the councils of the National Academy of Sciences, the National Academy of Engineering, and the Institute of Medicine. The members of the committee responsible for the report were chosen for their special competences and with regard for appropriate balance.

This is a report of work supported by Grant F49620-01-1-0269 between the U.S. Air Force and the National Academy of Sciences. Any opinions, findings, conclusions, or recommendations expressed in this publication are those of the author(s) and do not necessarily reflect the view of the organizations or agencies that provided support for the project.

International Standard Book Number 0-309-09175-6 (Book)
International Standard Book Number 0-309-53107-1 (PDF)

Limited copies are available from:

Air Force Science and Technology Board
National Research Council
500 Fifth Street, N.W.
Washington, DC 20001
(202) 334-3118

Additional copies are available from:

The National Academies Press
2101 Constitution Avenue, N.W.
Lockbox 285
Washington, DC 20055
(800) 624-6242 or (202) 334-3313
(in the Washington metropolitan area)
http://www.nap.edu

Printed in the United States of America

THE NATIONAL ACADEMIES

Advisers to the Nation on Science, Engineering, and Medicine

The **National Academy of Sciences** is a private, nonprofit, self-perpetuating society of distinguished scholars engaged in scientific and engineering research, dedicated to the furtherance of science and technology and to their use for the general welfare. Upon the authority of the charter granted to it by the Congress in 1863, the Academy has a mandate that requires it to advise the federal government on scientific and technical matters. Dr. Bruce M. Alberts is president of the National Academy of Sciences.

The **National Academy of Engineering** was established in 1964, under the charter of the National Academy of Sciences, as a parallel organization of outstanding engineers. It is autonomous in its administration and in the selection of its members, sharing with the National Academy of Sciences the responsibility for advising the federal government. The National Academy of Engineering also sponsors engineering programs aimed at meeting national needs, encourages education and research, and recognizes the superior achievements of engineers. Dr. Wm. A. Wulf is president of the National Academy of Engineering.

The **Institute of Medicine** was established in 1970 by the National Academy of Sciences to secure the services of eminent members of appropriate professions in the examination of policy matters pertaining to the health of the public. The Institute acts under the responsibility given to the National Academy of Sciences by its congressional charter to be an adviser to the federal government and, upon its own initiative, to identify issues of medical care, research, and education. Dr. Harvey V. Fineberg is president of the Institute of Medicine.

The **National Research Council** was organized by the National Academy of Sciences in 1916 to associate the broad community of science and technology with the Academy's purposes of furthering knowledge and advising the federal government. Functioning in accordance with general policies determined by the Academy, the Council has become the principal operating agency of both the National Academy of Sciences and the National Academy of Engineering in providing services to the government, the public, and the scientific and engineering communities. The Council is administered jointly by both Academies and the Institute of Medicine. Dr. Bruce M. Alberts and Dr. Wm. A. Wulf are chair and vice chair, respectively, of the National Research Council.

www.national-academies.org

COMMITTEE ON THE NATIONAL AEROSPACE INITIATIVE

EDSEL D. DUNFORD, *Chair*, TRW (retired)
DONALD J. KUTYNA, *Vice-Chair,* Loral Space and Communications, Colorado Springs
KEVIN G. BOWCUTT, The Boeing Company, Huntington Beach, California
KENNETH E. EICKMANN, University of Texas at Austin
WESLEY L. HARRIS, Massachusetts Institute of Technology, Cambridge
HANS G. HORNUNG, California Institute of Technology, Pasadena
KATHLEEN C. HOWELL, Purdue University, West Lafayette, Indiana
ERIC J. JUMPER, University of Notre Dame, Notre Dame, Indiana
IRA F. KUHN, JR., Directed Technologies, Arlington, Virginia
ANDREW J. MEADE, Rice University, Houston, Texas
CARL J. MEADE, Lockheed Martin Aeronautics, Palmdale, California
NEIL E. PATON, Liquidmetal Technologies, Lake Forest, California
RONALD F. PAULSON, Lockheed Martin Corporation, Bethesda, Maryland
FRED E. SAALFELD, National Defense University, Washington, D.C.
DONNA L. SHIRLEY, University of Oklahoma, Norman
PETER STAUDHAMMER, Northrop Grumman, Redondo Beach, California

Air Force Science and Technology Board Liaisons

ROBERT A. FUHRMAN, Lockheed Corporation (retired), Pebble Beach, California
ELI RESHOTKO, Case Western Reserve University (emeritus), Cleveland, Ohio

Staff

JAMES C. GARCIA, Study Director
LaNITA JONES, Project Assistant
DANIEL E.J. TALMAGE, JR., Research Associate
ANDREW WALTHER, Intern

AIR FORCE SCIENCE AND TECHNOLOGY BOARD

ROBERT A. FUHRMAN, *Chair*, Lockheed Corporation (retired), Pebble Beach, California
R. NOEL LONGUEMARE, *Vice-Chair*, Private Consultant, Ellicott City, Maryland
FRANK CAPPUCCIO, Lockheed Martin Corporation, Fort Worth, Texas
LYNN CONWAY, University of Michigan, Ann Arbor
LAWRENCE J. DELANEY, Titan Corporation, Arlington, Virginia
STEVEN D. DORFMAN, Hughes Electronics (retired), Los Angeles, California
EARL H. DOWELL, Duke University, Durham, North Carolina
DELORES M. ETTER, U.S. Naval Academy, Annapolis, Maryland
CHANDRA KUMAR N. PATEL, University of California at Los Angeles
RICHARD R. PAUL, The Boeing Company, Seattle, Washington
ROBERT F. RAGGIO, Dayton Aerospace, Inc., Dayton, Ohio
ELI RESHOTKO, Case Western Reserve University (emeritus), Cleveland, Ohio
LOURDES SALAMANCA-RIBA, University of Maryland, College Park
EUGENE L. TATTINI, Jet Propulsion Laboratory, Pasadena, California

Staff

MICHAEL A. CLARKE, Director
WILLIAM E. CAMPBELL, Administrative Officer
CHRIS JONES, Financial Associate
DEANNA P. SPARGER, Administrative Associate
DANIEL E.J. TALMAGE, JR., Research Associate

Preface

Since the end of the Cold War, the percentage of national resources devoted to aerospace has declined and graduation rates in science and engineering have declined as well. The goal of the National Aerospace Initiative (NAI), a partnership set up in 2001 between the Department of Defense (DoD) and the National Aeronautics and Space Administration (NASA), is to sustain U.S. leadership in aerospace in the coming decades. The initiative challenges the military services and agencies to accelerate development and demonstration milestones in selected areas to allow systems to be implemented earlier than they would otherwise have been.

BACKGROUND AND SCOPE OF STUDY

As the primary DoD participant in NAI, the Air Force became concerned about possible effects on its program and budget if NAI investment decisions followed a set of priorities different from those of the Air Force. For an independent assessment of the feasibility and operational relevance of NAI, the Air Force turned to the National Academies. In March 2003, the Deputy Assistant Secretary of the Air Force for Science, Technology, and Engineering requested a detailed study of NAI. The full statement of task is given in Box P-1. The study grant was awarded in mid-May 2003, after which the Committee on the National Aerospace Initiative was formed under the auspices of the National Research Council's (NRC's) Air Force Science and Technology Board (see Appendix A for short biographies of committee members). The first committee meeting was held in early August 2003. By agreement with the sponsor, the committee addressed two of the three NAI "pillars" (subject areas)—hypersonics and access to space—but did not attempt to comment on space technology.[1]

[1]It was agreed that the broad scope of the third NAI pillar—space technology—and the DoD security classification of much of the pertinent related information would limit the committee's approach. The first two NAI pillars—hypersonics and access to space—had narrower scopes and largely involved unclassified information, and recent budget proposals made these two pillars subjects of nearer-term concern.

Box P-1 Statement of Task

To assist the Department of Defense, the services and agencies, and NASA by providing an independent evaluation of the feasibility of achieving the science and technical goals as outlined in the National Aerospace Initiative, the National Academies, under the leadership of the Air Force Science and Technology Board, will form a committee to answer the following general questions concerning the NAI:

1. Is it technically feasible in the time frame laid out?
2. Is it financially feasible in the same time frame?
3. Is it operationally relevant?

In developing its answers, the committee will perform the following tasks:

- Examine information provided by DoD and NASA that defines, in broad terms, the goals for NAI to include enabling technologies needed to support the effort and the types of capabilities enabled.
- Evaluate the expected output from the science and technology implied by the NAI in terms of warfighter capability requirements.
- Make recommendations on the relevance of implied NAI S&T solutions to meeting these requirements as compared to other possible options. Assess impact on current service efforts to meet these capability needs.
- Baseline the current technology readiness of these requisite technologies and provide a committee estimate of associated technology development timelines. This estimate should take into account the professional opinion on how quickly relevant technologies can be matured.
- Identify and make recommendations for the technologies that should be emphasized over the next five to seven years to expedite overall roadmap accomplishment. The committee should consider two budget scenarios for the development of NAI timelines; one that recognizes the current constrained Air Force budget, assuming no additional NAI funds are allocated, and one that meets the optimal NAI development timelines as developed by the committee. Provide a rough order of magnitude estimate of the difference.
- Provide independent recommendations on specific efforts that could advance the areas of hypersonic propulsion, access to space, and space technology to meet warfighter needs over the next 20 years.
- Suggest initiatives required to ensure a more robust aerospace science and engineering workforce is available to meet these needs.

When the committee began this study, most committee members assumed they would be reviewing a clearly defined program with a strong management organization. In fact, the committee discovered that NAI included programs that predated the initiative and that the NAI executive office had only recently been staffed and was functioning as an advocate, facilitator, and data-sharing mechanism, with financial and management responsibility for the various programs remaining with the services and agencies.

STUDY APPROACH AND CONSTRAINTS

Over a 3-month period, the committee gathered data and information by meeting with persons involved in NAI planning, budgeting, and execution and by reviewing relevant reports and other documents. Appendix B lists presentations made to the committee by guest speakers.

Committee members met with the Director of Defense Research and Engineering (DDR&E) three times to receive information that was unclassified and cleared for public release, export-controlled information, and DoD classified information. The vice chair of the committee and the director of the Air Force Science and Technology Board, both with appropriate active security clearances, were briefed at a highly classified level. It was determined that the content of that briefing did not materially affect the findings and conclusions of this report. As requested by the Air Force, the committee's final report is unclassified; however, it is based on the understanding the committee received from all the information presented to it. The report does not (and could not) reflect information that was not presented to the committee.

During its first meeting, the committee divided itself into two main writing teams—one for hypersonics and one for access to space. Air-breathing hypersonics is an embryonic technology with considerable promise but no operational systems, while rocket-based vehicles have been operational as space launch or missile systems for 50 years. Because of the enormous difference in their operational maturity, the information presented to the committee differed substantially for the two topics. Discussion in this report of hypersonics and space access reflects these differences.

In general, the committee's approach to assessing NAI's technical feasibility was to analyze the main technical challenges to achieving NAI technical objectives and then decide whether NAI addresses those challenges. The committee did not attempt to predict whether all the challenges would be met. There are unknowns that despite DoD's and NASA's best efforts might not be resolved. NAI technical goals may be achievable and would certainly be useful if they were achieved; however, no one can guarantee that executing the best possible NAI plan will result in their achievement. The committee did its best to address technical feasibility separately from financial feasibility; however, in reality, the two are intertwined. NAI technical objectives cannot be achieved without money to pay for the needed research and technology development effort.

The inability to clearly determine NAI funding adversely affected the committee's ability to assess the financial feasibility of NAI. A clear understanding of NAI funding is also needed to consider current versus optimal budget scenarios and to provide related advice on NAI planning.

Estimating the investment required to develop technology is difficult under the most optimal conditions. Therefore, when even a rough estimate was beyond the scope of the study, the committee strove to evaluate what it could—namely, the relative utility of the technology area. An accurate and complete cost estimate by independent professionals who are expert in the practice should be completed as a follow-on to this study.

Finally, to assess the operational relevance of NAI, the committee looked for formal user requirements documents for NAI technologies or systems using NAI technologies. However, the committee did not base its conclusions solely on existing documents but rather sought indicators that such technologies could have a substantial payoff for the various military missions.

It was beyond the committee's ability to conduct an exhaustive review and comparison of all the options and alternatives for satisfying current warfighter requirements or providing future warfighting capabilities.

NASA'S NEW SPACE EXPLORATION MANDATE

On January 14, 2004, President Bush publicly announced "a new plan to explore space and extend a human presence across our solar system."[2] The President's plan called for developing and

[2]President Bush Announces New Vision for Space Exploration Program. Remarks by the President on U.S. Space Policy. NASA Headquarters. Washington, D.C. Speech available at http://www.whitehouse.gov/news/releases/2004/01/20040114-3.html. Last accessed on March 25, 2004.

testing a new crew exploration vehicle (CEV) by 2008, human missions to the Moon as early as 2014, and, later, human missions to Mars.

The President's plan was announced after the committee had completed its study and submitted its draft report for external peer review. It was clear to the committee and peer reviewers that the new NASA mandate could affect NAI as NASA's plans, programs, and resources shift toward new objectives. On February 9, 2004, the committee held a teleconference with Robert Shaw, Special Assistant to DDR&E, to discuss the likely outcome of the new mandate. Exactly how NAI will be affected is not yet clear; however, Mr. Shaw conjectured that some NAI schedule objectives might be significantly delayed. Technical objectives could change as well.

Despite the timing of the announcement and its uncertain consequences, the committee wanted this report to be as relevant as possible. In the limited time it had available, the committee reviewed the report and made revisions that it felt were reasonable. The committee found effects on NAI's access-to-space pillar easier to foresee than effects on its hypersonics pillar. Access to space is obviously relevant to development of the CEV and human missions to the Moon and Mars. What role hypersonics will play is not obvious at this time.

The committee advises readers of this report to keep in mind the reorientation now under way at NASA and the effects that this reorientation might have on the future of NAI.

ACKNOWLEDGMENTS

The members of the NRC study committee were highly motivated and intellectually curious and represented a wide range of academic, industrial, and military backgrounds. Because of a short schedule to cover such a complex subject, the meeting sessions were lengthy and the period of report drafting was abbreviated. In spite of this, every member of the committee willingly accepted his/her writing assignment, and many of them made site visits to organizations with programs in the subject areas.

The committee thanks the many organizations and guest speakers that provided excellent support to the committee. All the speakers were impressive and presented information to the committee that had a direct bearing on the study. From the high quality of the presentations, it was obvious that the speakers and others had spent many hours preparing. For the committee, this was time well spent. We hope that the speakers, their organizations, the committee's Air Force sponsor, and the ultimate readers of this report will agree.

Finally, the committee thanks the NRC staff members who supported the study. Primary among them were Mike Clarke, Jim Garcia, LaNita Jones, Daniel Talmage, and intern Andy Walther.

Edsel D. Dunford, *Chair*
Committee on the National Aerospace Initiative

Acknowledgment of Reviewers

This report has been reviewed in draft form by individuals chosen for their diverse perspectives and technical expertise, in accordance with procedures approved by the National Research Council's Report Review Committee. The purpose of this independent review is to provide candid and critical comments that will assist the institution in making its published report as sound as possible and to ensure that the report meets institutional standards for objectivity, evidence, and responsiveness to the study charge. The review comments and draft manuscript remain confidential to protect the integrity of the deliberative process. We wish to thank the following individuals for their review of this report:

Darrell R. Branscome, Science Applications International Corporation,
Yvonne C. Brill, Consultant,
Robert P. Caren, Lockheed Martin Corporation (retired),
Aloysius G. Casey, U.S. Air Force (retired),
Stewart E. Cranston, Veridian Engineering,
Werner J.A. Dahm, University of Michigan,
Delores M. Etter, U.S. Naval Academy,
Alexander H. Flax, Consultant,
Delma C. Freeman, National Aeronautics and Space Administration (retired),
George A. Paulikas, Aerospace Corporation (retired),
Todd I. Stewart, Ohio State University, and
John M. Swihart, Swihart Consulting, Inc.

Although the reviewers listed above have provided many constructive comments and suggestions, they were not asked to endorse the conclusions or recommendations nor did they see the final draft of the report before its release. The review of this report was overseen by Robert A. Frosch (NAE), Harvard University. Appointed by the NRC, he was responsible for making certain that an independent examination of this report was carried out in accordance with institutional procedures and that all review comments were carefully considered. Responsibility for the final content of this report rests entirely with the authoring committee and the institution.

Contents

Acronyms

ACC	Air Combat Command
AEDC	Arnold Engineering Development Center
AFB	Air Force Base
AFMC	Air Force Materiel Command
AFOSR	Air Force Office of Scientific Research
AFRL	Air Force Research Laboratory
AFROC	Air Force Requirements Oversight Council
AF SAB	Air Force Scientific Advisory Board
AFSPC	Air Force Space Command
AIA	Aerospace Industries Association
AIT	Atmospheric Interceptor Technology (program)
AMCOM	Aviation and Missile Command
AMRAAM	advanced medium-range air to air missile
APU	auxiliary power unit
AQR	Deputy Assistant Secretary of the Air Force for Science, Technology, and Engineering
ARRMD	affordable rapid response missile demonstrator
ASC	Aeronautical Systems Center
ASCI	Accelerated Strategic Computing Initiative
ASNRDA	Assistant to the Secretary of the Navy for Research, Development, and Acquisition
ASTP	Advanced Space Transportation Program
ATS	access-to-space
BAU	business as usual
BMDO	Ballistic Missile Defense Organization

C4ISR	command, control, communications, computing, intelligence, surveillance, and reconnaissance
CAV	common aero vehicle
CC	commander
CDR	critical design review
CFD	computational fluid dynamics
CFUSAI	Commission on the Future of the United States Aerospace Industry
CMC	ceramic matrix composite
CoDR	conceptual design review
CONOPS	concept of operations
CONUS	continental United States
CRRA	capability review and risk assessment
CSAF	Chief of Staff of the Air Force
CUBRC	Calspan-University of Buffalo Research Center, Inc.
CV	vice commander
DAKOTA	Design Analysis Kit for Optimization and Terascale Applications
DARPA	Defense Advanced Research Projects Agency
DCR	dual combustion ramjet
DDR&E	Director of Defense Research and Engineering
DES	discrete-eddy simulation
DMF	dry mass fraction
DoD	Department of Defense
DOE	Department of Energy
DSB	Defense Science Board
DSMC-NS	direct simulation Monte Carlo–Navier-Stokes
EELV	evolved, expendable launch vehicle
ERV	expendable rocket vehicle
FALCON	Force Application and Launch from CONUS (program)
FEM	finite element model
FLRS	future long-range strike
FRSC	fuel-rich staged combustion
FSD	full-scale development
FSW	friction stir welding
FY	fiscal year
FYDP	Future Years Defense Program
GASL	General Applied Science Laboratory
GDP	gross domestic product
GNC	guidance, navigation, and control
GOTChA	goals, objectives, technical challenges, and approaches
GPS	Global Positioning System
GRC	Glenn Research Center
GRST	Global Response Task Force
GSTF	Global Strike Task Force
GT	ground testing

H_2	diatomic hydrogen
HC	hydrocarbon
HCV	hypersonic cruise vehicle
HQ	headquarters
HS/H	high speed/hypersonics
HTHL	horizontal takeoff/horizontal landing
HyFly	Hypersonics Flight Demonstration (program)
HyTech	Hypersonics Technology (program)
IHPRPT	integrated high-payoff rocket propulsion technology
IHPTET	integrated high-performance turbine engine technology
IOC	initial operational capability
IP	integrated powerhead
IPD	integrated powerhead demonstrator
ISR	intelligence, surveillance, and reconnaissance
ISS	International Space Station
IVHM	integrated vehicle health management
JHU/APL	Johns Hopkins University/Applied Physics Laboratory
JROC	Joint Requirements Oversight Council
LaRC	Langley Research Center
LEO	low Earth orbit
LES	large-eddy simulation
LOx	liquid oxygen
LH_2	liquid hydrogen
LRS	long-range strike
MAJCOM	major command
MCH	methylcyclohexane
MDA	Missile Defense Agency
MDO	multidisciplinary design optimization
MIPCC	mass injection precompressor cooling
MIS	modular insertion stage
MMC	metal matrix composites
MNS	mission needs statement
MPV	MIPCC-powered vehicle
MSFC	Marshall Space Flight Center
NAI	National Aerospace Initiative
NASA	National Aeronautics and Space Administration
NASA HQ/MDepAA	Office of Space Flight Deputy Associate Administrator
NASA HQ/RAA	Office of Aeronautics Associate Administrator
NASP	National Aerospace Plane
NAVAIR	Naval Air Systems Command
NDAA	National Defense Authorization Act
NGLT	Next-Generation Launch Technology (program)
NIST	National Institute of Standards and Technology
NRC	National Research Council

NRO	National Reconnaissance Office
OML	outer mold line
OMS	orbital maneuvering system
ONR	Office of Naval Research
ORDs	operational requirements document
ORS	Operationally Responsive Spacelift
ORSC	oxidizer-rich staged combustion
ORU	orbital replacement unit
OSD	Office of the Secretary of Defense
OSP	orbital space plane
OSTP	Office of Science and Technology Policy
OTV	orbit transfer vehicle
P&W	Pratt & Whitney
PBR	President's budget request
PDR	preliminary design review
PGS	Prompt Global Strike
PLIF	planar laser-induced fluorescence
PRD	program requirements document
R&D	research and development
RAA	regional airline association
RANS	Reynolds-averaged Navier-Stokes
RASCAL	Responsive Access, Small Cargo, Affordable Launch (program)
RATTLRS	Revolutionary Approach to Time-Critical Long-Range Strike (program)
RBCC	rocket-based combined cycle
RCS	reaction control system
RDT&E	research, development, test, and evaluation
RFI	Resource Conservation and Recovery Act facility investigation
RLV	reusable launch vehicle
RP	rocket propellant
RTA	Revolutionary Turbine Accelerator (program)
S&E	science and engineering
S&T	science and technology
SAALT	Secretary of the Army for Acquisition, Logistics, and Technology
SAF	Secretary of the Air Force
SBR	space-based radar
SC	Space Control
SDB	small-diameter bomb
SECAF	Secretary of the Air Force
SECDEF	Secretary of Defense
SED	Single-Engine Demonstrator (program)
SJ	scramjet
SLV	small launch vehicle
SMC	Space and Missile Systems Center

SMV	space maneuvering vehicle
SOA	state of the art
SOV	space operations vehicle
SSC	Stennis Space Center
SSTO	single stage to orbit
STEM	space, technology, engineering, and mathematics
STS	Space Transportation System (shuttles)
TBBC	turbine-based combination cycle
TCT	time-critical target
TDRSS	tracking and data relay satellite system
TEO	technology executive officer
TJ	turbojet
TOA	total obligational authority
TPS	thermal protection system
TRL	technology readiness level
TSTO	two stage to orbit
URETI	university research, engineering, and technology institute
USAF	U.S. Air Force
USECAF	Under Secretary of the Air Force
USMC	U.S. Marine Corps
USSTRATCOM	U.S. Strategic Command
V&V	validation and verification
VAATE	Versatile Affordable Advanced Turbine Engines (program)
VLS	vertical launch system
VMC	vehicle management computer
VMS	vehicle management system
WMD	weapons of mass destruction

Executive Summary

THE BOTTOM LINE

The committee believes that the National Aerospace Initiative (NAI) is an effective instrument for assisting the Department of Defense (DoD) and the National Aeronautics and Space Administration (NASA) in their pursuit of technologies for our nation's future military systems and its future space launch needs, both crewed and uncrewed. The NAI goals directly address issues of national concern. NAI has fostered collaboration between the DoD and NASA in research and technology areas—hypersonics, access to space, and space technology—that can benefit both organizations. It has increased the visibility of the two agencies' related research and technology development efforts and has made it easier for them to work together to achieve overlapping objectives. It has increased the probability of achieving synergistic results. It offers opportunities to achieve efficiencies that would not be possible if each agency were to work in isolation. The committee recommends that DoD and NASA, through NAI, continue to advocate, communicate, and facilitate the nation's endeavors in hypersonics, access to space, and space technology.

The stated mission of NAI is to "ensure America's aerospace leadership with an integrated, capability-focused, national approach that enables high speed/hypersonics flight; affordable, responsive, safe, reliable access to and from space; and in-space operation by developing, maturing, demonstrating, and transitioning transformational aerospace technologies."[1] For hypersonics and space access, NAI goals are stated as follows:

- *Hypersonics.* Flight demonstrate increasing Mach number capability each year, reaching Mach 12 by 2012.
- *Space access.* Demonstrate technologies that will dramatically increase space access and reliability while decreasing costs.[2]

[1]DDR&E. 2004. National Aerospace Initiative. Mission Statement. Available at http://www.dod.mil/ddre/nai/mission.html. Last accessed on April 04, 2004.

[2]DDR&E. 2003. National Aerospace Initiative. Sustaining American Aerospace Initiative. Available at http://www.dod.mil/ddre/nai/nai_brochure1.pdf. Accessed on December 23, 2003.

The study committee was asked to answer three general questions—Is NAI technically feasible in the time frame laid out? Is it financially feasible in the same time frame? Is it operationally relevant? The committee's answers are presented in the paragraphs that follow.

Concerning the technical feasibility of hypersonics, the NAI roadmap for integrated high speed/hypersonics and space access ground and flight demonstration outlines a series of development and demonstration programs resulting in a Mach 12 air-breathing capability in the 2014 time frame. While the committee believes that Mach 12 air-breathing vehicles may be technically feasible in that time frame, it recommends a more comprehensive approach addressing all requisite activities, from fundamental research to critical technology development to flight demonstration.

The NAI phased approach to space access with rocket propulsion envisions that technology investments will result in increasingly ambitious potential system payoffs by 2008 and 2015. The quantified payoffs include short turnaround time; high sortie numbers for airframe, propulsion, and systems; low marginal sortie cost; high reliability; and improved payload performance. The committee believes strongly in the general goal of demonstrating technologies to dramatically increase space access and reliability while decreasing cost but does not believe that all the payoffs will be available in the time frames suggested by the NAI.

Concerning financial feasibility, the committee believes that both pillars are underfunded in relation to current NAI planning. It believes that near-term NAI funding for the air-breathing hypersonics pillar might suffice for a significant critical technologies program that could support near-term warfighting applications such as missiles. However, sharply higher budgets will be required to achieve the currently stated, long-term NAI objective of air-breathing hypersonic access to space. The access-to-space pillar faces a similar funding issue. The NAI envisions a multiphase demonstration program with increasingly capable reusable rockets available in 2008 and 2015. The development of these vehicles is not supported by current budgets. Clearly, neither the goals of NAI nor the needs of the military services can be met without significant additional funding.

Finally, the committee found that NAI is operationally relevant. All the DoD operational commands contacted by the committee believe that NAI technologies and capabilities, if realized, could support, to a degree, their stated capability goals and missions, such as Prompt Global Strike, Global Missile Defense, and Operationally Responsive Spacelift. However, non-NAI approaches can also support these missions, and the operational commands recognize that many NAI technologies have yet to be developed and proven.

The committee strongly agrees with the operational commands, which, while they support the capabilities offered by NAI, believe that NAI must be balanced with other research and technology development efforts, priorities, and investments to ensure proper trades between current, near-term, and future combat capability.

STUDY TASKS

As described in the preface, this study was undertaken in response to a request by the U.S. Air Force that the National Research Council (NRC) of the National Academies provide an independent evaluation of the feasibility of achieving the science and technical goals outlined by the NAI. To conduct the study, the NRC appointed the Committee on the National Aerospace Initiative under the auspices of the Air Force Science and Technology Board.

To answer the three general questions the study committee was asked—Is NAI technically feasible in the time frame laid out? Is it financially feasible in the same time frame? Is it operationally relevant?—the committee was asked to perform several tasks, including evaluating NAI in terms of warfighter capabilities and baselining the readiness of NAI technologies. The committee was also asked to recommend technologies that should be emphasized over the next 5 to 7 years as well as specific efforts to advance hypersonics and access to space over the next 20 years. In addition, the committee was asked to consider two Air Force budget scenarios—one assuming that no additional NAI funds are allocated and one corresponding to optimal NAI development timelines. Finally, the committee was asked to suggest initiatives to ensure a more robust aerospace science and engineering workforce.

THE NATIONAL AEROSPACE INITIATIVE

The National Aerospace Initiative is a joint technology initiative begun in 2001 by the DoD and NASA. The goals of NAI are to renew American aerospace leadership; push the space frontier with breakthrough aerospace technologies; revitalize the U.S. aerospace industry; stimulate science and engineering education; and enhance U.S. security, economy, and quality of life. The initiative focuses on science and technology advances in three areas, or "pillars"—high speed/hypersonics flight, access to space, and space technologies. The high speed/hypersonics flight goal is to demonstrate Mach 12 by 2012. The access-to-space goal is to demonstrate technologies to dramatically increase space access and reliability while decreasing costs. Leveraging the full potential of space is the goal of the space technology pillar, an area that, by agreement with the sponsor, the committee did not address.

Air-breathing hypersonic vehicles have potential application as missiles, cruise missiles, long-range strike aircraft, and/or space launchers. Technology development and demonstration are required to mature critical hypersonic technologies to the point where a decision could be made to fund any or all of these applications. The technical challenges of air-breathing hypersonics increase dramatically as the speed of the vehicle increases, through scramjet speeds that could reach Mach 4 to Mach 14. The military departments and NASA have ongoing technology development and demonstration programs in air-breathing hypersonics. The Army and Navy programs are directed toward missile applications. The Air Force is interested in nearer-term missile applications and is also working with NASA to pursue the longer-term possibilities of reusable launch vehicles for space access.

During the past 45 years, space access capability has been developed for low-launch-rate applications from low Earth orbits to beyond the solar system. Rocket-based vehicles have operated as missiles, long-range strike missiles, or space launchers for 50 years. The Air Force and NASA are both investing in technology for rocket-based, reusable launch vehicles. While their final systems may be very different, they share many common technologies, for which NAI is facilitating related collaboration. The Air Force has credible emerging needs for a rapid rate, operationally responsive spacelift capability that might utilize a reusable rocket-powered vehicle. NASA needs a replacement for its Space Transportation System (the space shuttles) that might be implemented in the next decade using reusable, rocket-based propulsion. Both DoD and NASA are planning new rocket-based systems for the near term but are at the same time working on capabilities in air-breathing hypersonics for possible application to launch vehicles in about two decades.

Advances in space technology are desired to provide national security decision makers with the most current and complete information made available through on-demand intelligence, surveillance, and reconnaissance. In addition, space technology advances are desired to benefit other warfighter mission areas such as navigation, weather, communications, missile warning, space control, and force application. Satellite systems using advanced space technologies are seen by NAI participants as a vital element of space, air, and ground systems that are so well networked and integrated that they provide revolutionary capabilities. The nonmilitary benefits of advanced space technology include the scientific study of Earth, leading to improved capabilities for predicting climate, weather, and natural hazards.

During 2002, workshops and meetings were held to further develop NAI definitions, goals, and plans. Planning teams were formed around the three NAI pillars. They included participants from the three military departments, NASA, and staff of the Director of Defense Research and Engineering (DDR&E). The teams used a structured process to go from the high-level NAI goal statements to project-level technology roadmaps. The process combined layered analysis (goals were analyzed to determine objectives, which were analyzed to determine technical challenges, and so on) and planning (projects were identified and roadmaps developed to address the challenges). By the time of this study, NAI participants had undergone the first iteration of this process and had developed initial roadmaps that included project-level details. The initial roadmaps included many DoD and NASA NAI-related projects that had already been under way or planned when NAI was established.

The NAI executive office, staffed by DoD and NASA personnel, acts as an advocate and facilitates collaboration in the development of goals, plans, and roadmaps for the three pillars.

Late in the budget cycle, DDR&E succeeded in having new NAI funding (in addition to that for existing or planned projects) included in the President's budget request (PBR) for fiscal year (FY) 2004. No new NAI funding was included in the PBR for FY 2005 and beyond; however, DDR&E expected the FY 2005 PBR to add new out-year funding.

THE NAI ROADMAPPING PROCESS

The committee believes that DoD and NASA have made a positive start; however, it also noted some weaknesses in what has been done so far. It was not clear to the committee that the process used to develop the NAI roadmaps was effective in defining a comprehensive and compelling technology development program that would mature all of the critical enabling technologies in time to meet the various NAI schedule goals, one of them being a milestone decision in 2018 on hypersonic access to space. For example, NAI brought together essentially all existing U.S. programs and attempted to use them to lay out a high speed/hypersonics technology roadmap. However, this roadmap is more a collection of existing programs than a logical and complete plan to achieve military strike, global reach, and space access objectives. Each of the collected programs has elements that address some portion of the critical technologies, and each program, owing to its very existence, has a funding line. The collective funds, if properly applied, perhaps are sufficient in the near term for a significant critical technologies program; however, as configured, it is not clear that the collected programs cover all critical hypersonic technologies. The low level of basic and applied research in the NAI plan is also conspicuous.

The committee recommends that, starting with a defined and articulated vision, DoD and NASA use a top-down process based on sound system engineering principles to determine the objectives, technical challenges, and enabling technologies and to plan the fundamental research, technology development, ground testing, and flight demonstrations required to mature the enabling technologies to levels sufficient for application. The result should be a comprehensive, integrated roadmap that ensures all technologies are sufficiently matured to support the multitude of decision milestones scheduled during the NAI time frame of interest. This roadmap may include preexisting projects; however, the roadmapping process should closely examine how each project contributes to achieving NAI goals and assess the trade-offs with other needed technology efforts. The roadmap should include detailed plans for fundamental research and clearly defined exit criteria for each of the critical technologies. The committee recommends that DoD and NASA clearly communicate the plan, once it is complete, to decision makers and stakeholders, including the public.

The committee recommends that DoD and NASA complete an end-to-end cost estimate for the top-down program through the period of interest and then work to secure funding commitments consistent with this cost. This plan and its funding estimates should then be exposed to competition from other DoD and NASA requirements to see if they are realistic. For example, if the national plan to complete and begin initial operational capability of a multibillion-dollar new and revolutionary two-stage-to-orbit, reusable launch vehicle is achieved in 2015-2018, how likely is it that at the same time Congress and the nation will be willing to approve an even more expensive follow-on air-breathing hypersonic launcher?

To help assess the realism of the resulting roadmaps, the committee recommends that DoD and NASA set up their planned NAI advisory panels, steering groups, and revolutionary concepts panels to review the NAI program on a continuing basis. The committee believes that the progress NAI has made in facilitating coordination of activities among participants would benefit from periodic oversight by independent groups of experts.

HYPERSONICS TECHNOLOGIES

The committee identified four critical enabling technologies for air-breathing hypersonic flight that must be matured: air-breathing propulsion and flight test; materials, thermal protection systems, and structures; integrated vehicle design and multidisciplinary optimization; and integrated ground test and numerical simulation and analysis.

Propulsion is foremost among the critical technologies that will enable air-breathing hypersonic flight, both for the dual-mode ramjet/scramjet engines that will achieve hypersonic speeds and for the lower speed engines that will accelerate the ramjet/scramjet to takeover speeds (typically Mach 3-4). Several challenges exist for scramjet engines, among them performance and operability across a broad range of operating speeds, especially above Mach 8; management of the extreme thermal environment encountered by such engines; and durability of engine structures and systems for long life and low maintenance. Challenges for low-speed engines include thrust-to-weight ratio (i.e., high thrust at low weight while maintaining high efficiency), thermal management, and integration with both the airframe and the high-speed propulsion system.

The committee focused on three important areas of materials technology: thermal protection systems, actively cooled combustor panels, and cryogenic tanks. The silica foam tiles used for thermal protection on the space shuttle are fragile and require extensive maintenance between missions; they are not suitable where rapid response is a requirement. More rugged thermal protection systems are needed that are passive (uncooled) or active (cooled). In the combustor section of a hypersonic propulsion system, active cooling is necessary, even at modest Mach numbers. Although extensive testing of various candidate designs has been accomplished, no single solution to all the operational requirements has emerged. Cryogenic tanks have received considerable attention in recent years, but no fully reusable tank has been adequately tested in the flight environment. The remaining step is to produce and flight test a full-scale, reusable tank through many cycles.

Air-breathing hypersonic vehicles will consist of highly integrated systems that will require multidisciplinary design optimization to obtain robust vehicle designs that satisfy all constraints. At hypersonic Mach numbers, much of the airframe must act as the inlet and nozzle for the propulsion system. The shape of the vehicle will determine the vehicle structure, the type of integrated thermal protection system and its material, the control system, the flight mechanics, and the flight trajectory. The flight trajectory, in turn, will determine the aerodynamic heating loads, which will influence vehicle aeroelastic behavior, performance, and empty weight. The empty weight of the vehicle is directly related to the structural shape, and the vehicle airframe will also affect the fuel and payload volume since, unlike conventional aircraft, the greater part of a hypersonic vehicle's volume must accommodate fuel. Several capabilities needed for multidisciplinary design optimization are not yet mature.

Integrated ground test and numerical simulation and analysis is another critical and enabling technology area that must be matured. The number of parameters that must be simulated in a vehicle development program is increased substantially when the Mach number goes from 3 to 8 or more. Computational techniques, while largely successful in many low-Mach-number flow applications, suffer severely at hypersonic speeds, yet, owing to test facility limitations, they are essential for supplementing ground testing. Ground test facilities that can provide longer test times (seconds versus milliseconds) and more diagnostics are needed. Though a number of facilities exist for hypersonic and hypervelocity testing, they must be enhanced, along with diagnostics. One or more new facilities will also probably be required to meet the full needs of hypersonic system development.

Over the next 5 to 7 years, the committee recommends that, through NAI, DoD and NASA should continue to support programs for developing and flight testing small-scale hydrogen and hydrocarbon engines and should begin to develop a mid-scale scramjet engine and to demonstrate scramjet operation at the highest anticipated operational speeds (approximately Mach 14). For the long term, along the path to meeting the objective of a full-scale, air-breathing, hypersonic launch vehicle, the committee recommends that engine development proceed by increasing engine scale in at least three steps: small scale, medium scale, and large scale. The committee believes that DoD and NASA need to better coordinate and agree on their readiness definitions and assessments for high-temperature materials and that a reinvigorated basic and applied research program is needed to push these materials along. The committee believes that development of multidisciplinary design optimization specific to hypersonic vehicle design is long term and will require several more years of adequate, sustained funding. As for combined research in ground testing and computation, the committee believes that the DoD and NASA NAI roadmap should include benchmarking experiments at different facilities; external flow testing at the correct enthalpy and covering the parameter space in combination with application of validated tools; ground testing of missile-scale engines at all designed-for operational speeds; subscale testing of engines for hypersonic vehicles larger than missiles; and a 10-year research program on hypersonic flows, emphasizing high-enthalpy effects.

ACCESS-TO-SPACE TECHNOLOGIES

The committee studied a variety of inputs and reference sources pertaining to the NAI access-to-space pillar. Although NAI is chartered as a technology development effort, the access-to-space pillar goals are expressed in terms of future operational system characteristics. This approach allows NAI to cast its net across a broad range of possible contributing technologies but necessitates the translation from system characteristics to technology objectives—a process open to some interpretation. Furthermore, without a clear understanding of the eventual system configuration, the specific system characteristics specified in the three-phase pillar could be difficult to justify and might result in expending resources on low-payoff technologies.

That said, the committee found that many areas of research require attention under the NAI access-to-space pillar. Near-term and long-term technologies must be pursued, with appropriate weighting between the two. The technologies most likely to contribute to achieving the near-term goals of NAI are advanced materials for use in propulsion and thermal protection systems; integrated structures; electrical/hydraulic power generation and management technologies; software transportability; and error-free software generation and verification. Furthermore, the development of computational analysis tools and methodologies should be emphasized—especially when coupled to test analysis and ground test facilities.

Propulsion research work should focus on all technologies contributing to engine reusability and reliability, including (but not restricted to) the development of high-strength, liquid-oxygen-compatible materials and new engine materials that can support combustion of both hydrocarbon and hydrogen fuels. In addition to research on engine materials noted above, generic materials research should be focused on technologies contributing to reusability, such as lightweight thermal protection materials, and structural materials that would be useful in reducing the dry weight of highly integrated airframe designs. Durability should be a prime consideration in the development of these materials. Additional materials research should focus on reusable propellant tanks. The committee recommends that DoD and NASA support basic and applied materials research programs and their application to highly integrated structures.

Both DoD and NASA will benefit by leveraging the work being done by the Department of Energy for electrical power management and control. In this general area, advances in intelligent sensors and component thermal control are ready for attention from the NAI. Multiple programs supported by DoD and other agencies could also be applicable.

Software is one of the most demanding and expensive aspects of any modern aerospace vehicle. Delays in software development ripple through the entire development process and significantly increase the cost of the overall project. The committee recommends that DoD and NASA, through NAI, support a robust research effort devoted to lowering the costs of aerospace software production. This research should concentrate on the safety-critical nature of aerospace software.

The coupling of high-performance computing (numerical analysis) with the emerging capabilities of ground test facilities (analysis tools and methodologies) appears to show great promise of reducing the cost of vehicle development. Additionally, advances in vehicle health management technologies would pay dividends in safety, in engine life extension, and in reducing the scheduled maintenance burden. Although these technologies are often considered outside the core of traditional aerospace research, their advancement is a legitimate goal and should receive appropriate attention from NAI.

The committee recommends that DoD and NASA develop time-phased, reusable, rocket-based flight demonstration programs to move near-term, unproven technologies through flight test; specify and disseminate the technology readiness levels and specific exit criteria necessary to support operational decision points; ensure that research is directed toward obtaining the specified data and that the demonstrations—both flight and ground—are structured to obtain the required information/data; and concentrate on technologies that contribute to reusability.

AEROSPACE WORKFORCE

The goals of NAI—renewing American aerospace leadership, advancing aerospace technologies, revitalizing the aerospace industry, stimulating science and engineering education, and enhancing U.S security and the economy—are both affected by and concerned with the state of the U.S. science and engineering aerospace workforce. The ability to achieve NAI technical goals depends on this workforce. This workforce, in turn, can be enhanced through pursuit of these goals.

Fueled by what appear to be negative indicators, concern about the state of the U.S. aerospace workforce has been the subject of much discussion in recent years. These indicators show reductions in the number of persons employed in the aerospace industry, the decreasing U.S. share of the world aerospace market, the increasing average age of aerospace workers, decreasing undergraduate and graduate enrollment in science and engineering, and declining mathematics and science performance of U.S. students relative to that of students from other countries. If there is a direct relationship between the state of the aerospace workforce and national security, as some have argued, then one might also argue that national security is threatened. NAI goals explicitly address this concern.

The committee found that NAI participants are focused mainly on achieving the initiative's technical goals. It is mainly through the pursuit of these goals that the NAI affects the aerospace workforce; however, the committee also found some NAI efforts specifically aimed at stimulating aerospace education and workforce development. Two university research, engineering, and technology institutes (URETIs) have been jointly funded by DoD and NASA and included in NAI. They perform research in reusable launch vehicles and aerospace propulsion and power by bringing together students, faculty, NASA and DoD researchers, industry, and facilities. Student research symposia, scholarships and fellowships, K-12 outreach, outreach to women and underrepresented minorities, and researcher exchange are included. The NAI-affiliated URETIs are funded at about $3 million per year.

The committee recognizes that NAI alone cannot allay all concerns about the U.S. aerospace workforce; however, it believes that NAI can assist in a national effort to address those concerns. The committee believes that efforts to achieve NAI technical objectives can help the aerospace workforce by providing the most important component—stable and predictable funding. Despite finding several drawbacks to the current URETI model that should be addressed, the committee believes that the NAI URETIs are stimulating aerospace education and the workforce. The committee believes that pursuit of NAI's workforce-related goals can and should be strengthened. The committee recommends that DoD and NASA establish an aerospace workforce pillar along with the existing hypersonics, access-to-space, and space technology pillars.

1

The National Aerospace Initiative

INTRODUCTION

In 2001, the Director of Defense Research and Engineering (DDR&E) started a joint Department of Defense (DoD)–National Aeronautics and Space Administration (NASA) technology initiative called the National Aerospace Initiative (NAI). The stated mission of NAI was to "ensure America's aerospace leadership with an integrated, capability-focused, national approach that enables high speed/hypersonics flight; affordable, responsive, safe, reliable access to and from space; and in-space operation by developing, maturing, demonstrating, and transitioning transformational aerospace technologies" (Richman, 2003a).

NASA describes NAI as a "twenty-five year national technology plan to mature key technologies for NASA and Department of Defense needs" (Rogacki, 2003). Common technologies that both agencies need to accomplish their respective mission objectives include long-life rocket engines; combined-cycle propulsion; ram/scramjets; long-life, lightweight airframes and tanks; durable thermal protection systems; all-electric subsystems; and rapid-turnaround ground and flight operations (Rogacki, 2003). NAI is an effort by DoD and NASA to partner and cooperate to develop these technologies that both agencies need.

According to DDR&E, NAI is a technology initiative, not a DoD or NASA system development or acquisition program (Richman, 2003a). By DDR&E and NASA definition, the three aerospace areas, or "pillars," encompassed by NAI are hypersonics, access to space, and space technologies. A widely used illustration of the NAI technology framework is shown in Figure 1-1. Each area includes a collection of science and technology (S&T) projects, many or most of which were already planned or ongoing before NAI was started.[1] The three areas have some overlaps; however, they can also stand largely by themselves.

---

[1]A few of these are the following (more NAI projects are described in Chapters 2 and 3 and Appendixes C and D):

- Army Hypersonic Missile Technology,
- Navy Hypersonics Flight Demonstrator Program (HyFly),
- Air Force Single-Engine Demonstrator,
- The Defense Advanced Research Projects Agency (DARPA) Responsive Access, Small Cargo, Affordable Launch (RASCAL), and
- NASA Next-Generation Launch Technology (NGLT).

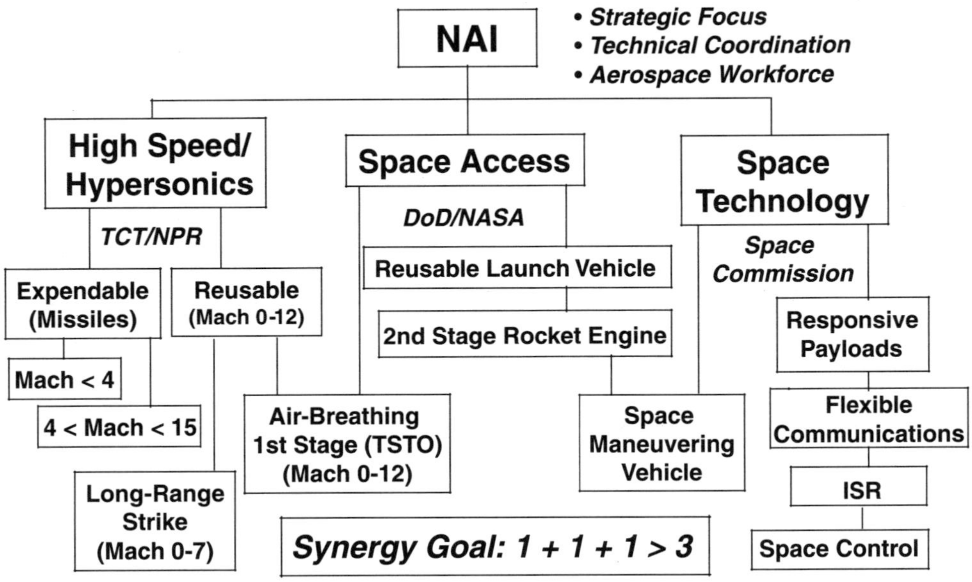

FIGURE 1-1 National Aerospace Initiative technology framework. ISR, intelligence, surveillance, and reconnaissance. SOURCE: Sega, 2003.

Examples of technologies and systems in each of the three pillars are shown in the figure. Implicit in the partnership between DoD and NASA is the goal of increasing the synergy and efficiency of the agencies' combined programs in the three NAI areas.

GOALS, PLANNING APPROACH, AND FUNDING

The high-level, national goals of the NAI include renewing American aerospace leadership in the 21st century, pushing the space frontier further and faster with breakthrough aerospace technologies, revitalizing our critical aerospace industry, stimulating science and engineering in our classrooms, and enhancing our security, economy, and quality of life (DDR&E, 2003).

NAI goals for each pillar are stated as follows (DDR&E, 2003):

- *Hypersonics*. Flight demonstrate increasing Mach number each year, reaching Mach 12 by 2012.
- *Space access*. Demonstrate technologies to dramatically increase space access and reliability while decreasing costs.
- *Space technology*. Leverage the full potential of space.

An NAI executive office with a director, leads drawn from both NASA and the DoD for the three pillars, and a small support staff have been established (Figure 1-2). The executive office acts as an advocate, facilitates collaboration, and develops goals, plans, and roadmaps in all three technology areas. The concept shown in Figure 1-2 is a work in progress; it has not yet been formally approved.

To go from the high-level goal statements for the initiative as a whole to project-level technology roadmaps, planning teams that included participants from the three military departments, NASA, and DDR&E staff were formed around the three NAI pillars (Richman, 2003a). The NAI

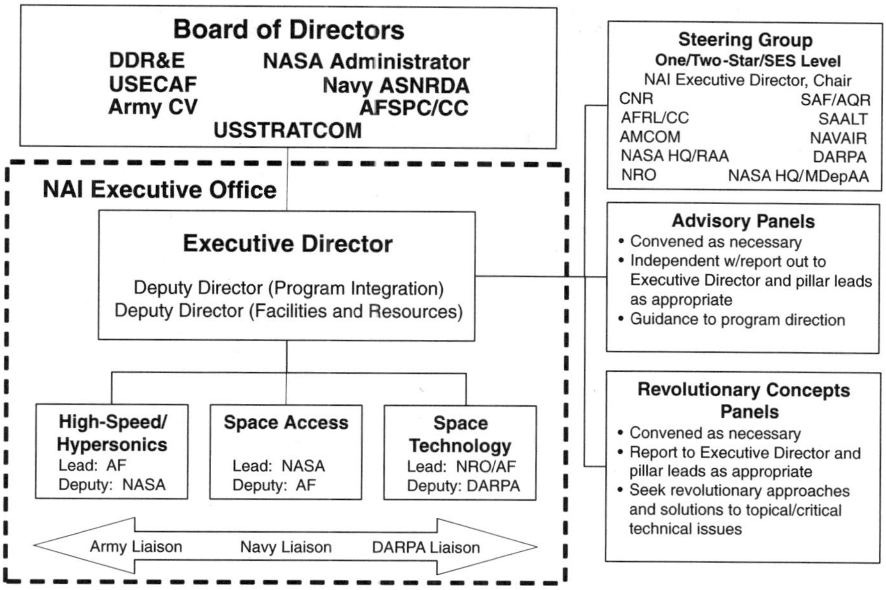

FIGURE 1-2 NAI executive office (draft). SOURCE: Sega, 2003.

planning teams that met during 2002 used a process they called GOTChA (goals, objectives, technical challenges, and approaches). The GOTChA process combines layered analysis (goals are analyzed to determine objectives, which are analyzed to determine technical challenges, and so on) and planning (projects are identified and roadmaps developed to address the challenges). Figure 1-3 illustrates the approach used by the planning teams. The GOTChA charts mentioned in Figure 1-3

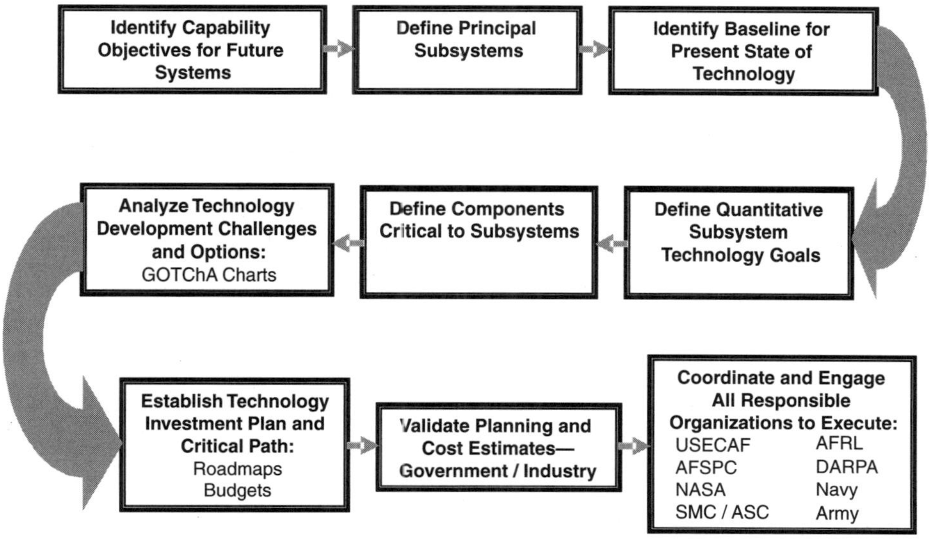

FIGURE 1-3 Technical analysis and planning approach used by NAI planning teams. SOURCE: Liston et al., 2003.

TABLE 1-1 Air Force NAI-Specific and Related Funding (thousand dollars)

Pillar	FY04 NAI Plus-Up Only	FY04[a]	FY05	FY06	FY07	FY08	FY09
Hypersonics	111,292	126,941	28,729	52,011	61,653	64,260	51,100
Space access	25,708	114,035	98,990	108,659	118,571	118,562	110,261
Space technology	24,500	137,415	115,266	114,714	122,961	164,595	165,103
Total	161,500	378,391	242,985	275,384	303,185	347,417	326,464

[a] FY 2004 amounts include plus-up.
SOURCE: Blackhurst, 2003.

show the flow downward from goals, through objectives, technical challenges, and approaches, to specific projects.[2] During that time, the military departments and DDR&E worked on their science and technology (S&T) program budget requests to be submitted to Congress during the winter of 2003 as part of the President's budget request (PBR) for fiscal year (FY) 2004.

Programs to be included in the NAI are technology development and demonstration activities, both new and ongoing before the initiative was set up. Responsibility for the funding and direction of these activities remains within the individual military departments and the two lead agencies, and they retain the right to establish budgets and redirect funds. The NAI, however, may use the appropriation process to obtain additional funds earmarked for specific technologies or demonstrations.

Late in the budget cycle, DDR&E succeeded in having new NAI funding (in addition to that allocated for projects already under way or planned) included in the President's budget request (PBR) for FY 2004. The PBR contained new NAI funding for FY 2004, not for FY 2005 or beyond (Richman, 2003b). Table 1-1 shows the plus-up Air Force NAI funding contained in the FY 2004 PBR.

The committee was unable to obtain data similar to those in Table 1-1 for the other NAI participants (other military departments, defense agencies, NASA, or NAI as a whole) and so could not obtain a clear, comprehensive picture of NAI funding. Although bits and pieces of funding data appeared in some of the presentations, the committee could not clearly determine what is being spent or is planned to be spent on NAI. Thus, it lacked the information it needed to assess NAI's financial feasibility.

RELEVANCE TO OPERATIONAL CAPABILITY REQUIREMENTS

The committee believes that any assessment of the operational relevance of the technologies and capabilities resulting from NAI must be based on their ability to contribute to the accomplishment of the operational needs of NASA and the military services. Those needs are best articulated by the organizations themselves and are contained in the military services' operational requirements documents (ORDs) and mission needs statements (MNS), and in NASA's program requirements documents (PRDs).

However, it is important to note that rarely is a research program established solely to solve some specific goal or objective. Instead, research programs typically lay the groundwork for and/or

[2]The NAI executive office showed the committee some sample GOTChA charts; however, due to concerns about the public releasability of some of the content (a mix of export-controlled, For Official Use Only, and draft information), it did not show the committee all or even a major portion of the GOTChA charts, of which there are hundreds.

assist in the development of technologies that promise to improve operational capabilities, including warfighting, access to space, or some other general goal or objective. As the Air Force pointed out, ". . . for transformational technologies there has never been a formal requirement for that technology" (AFRL, 2003a). When the Wright brothers developed the airplane and forever transformed human life, it was not in response to a requirement for such a vehicle. Much the same can be said for the evolution of the airplane into a combat-capable system and the development of precision guided munitions, stealth technology, and the Global Positioning System (GPS). Common to all of these transformational technologies was not a formal requirement but, rather, indicators that they could have a strong potential payoff. Based on information supplied by the services and NASA to the committee, this appears to be the case for the technologies and capabilities contained in the NAI.

To ensure that it properly understood assigned missions and the operational needs they drive, the committee contacted NASA, the Air Force Space Command, the U.S. Strategic Command, and the Air Combat Command, as well as representatives of the U.S. Navy and the U.S. Army. Each organization was asked to prepare a talking paper on NAI's operational relevance to its goals and missions. In addition, the Air Force Research Laboratory (AFRL) was asked to justify its NAI-related research. AFRL explained that its thrusts, priorities, and funding requirements are developed in response to current and future operational needs of the Air Force. The committee was particularly interested in what stated operational needs or requirements had driven the ongoing research and how those needs had contributed to the establishment of research goals and priorities for NAI.

The documents received from the U.S. military organizations and NASA as a result of these inquiries are summarized below.

Summary of Stakeholder Views

U.S. Air Force

The U.S. Air Force Posture Statement 2003 and America's Air Force Vision (AFRL, 2003b) describe the Air Force vision of Global Vigilance, Global Reach, and Global Power and the concepts of operations (CONOPS) under which the Air Force will operate. The objective is fourfold:

- To transform the Air Force into a capabilities-focused expeditionary air and space force.
- To make warfighting effects the drivers for everything the Air Force does.
- To provide necessary resources expeditiously to warfighters.
- To guide planning, programming, and requirements reform.

The six CONOPS developed in support of these objectives include Global Strike, Homeland Security, Global Mobility, Global Response, Nuclear Response, and Space and C4ISR (see Figure 1-4).

To understand how the technologies and capabilities contained in NAI support these CONOPS, it is valuable to look at how they are defined in the posture statement (AFRL, 2003a).

- *Global Strike Task Force (GSTF).* Employs joint power-projection capabilities to engage anti-access and high-value targets, gain access to denied battlespace, and maintain battlespace access for all required joint/coalition follow-on operations.
- *Global Response Task Force (GRTF).* Combines intelligence and strike systems to attack fleeting or emergent, high-value or high-risk targets by surgically applying air and space power in a narrow window of opportunity, anywhere on the globe, within hours.
- *Space and C4ISR Task Force.* Harnesses horizontal integration of manned, unmanned and

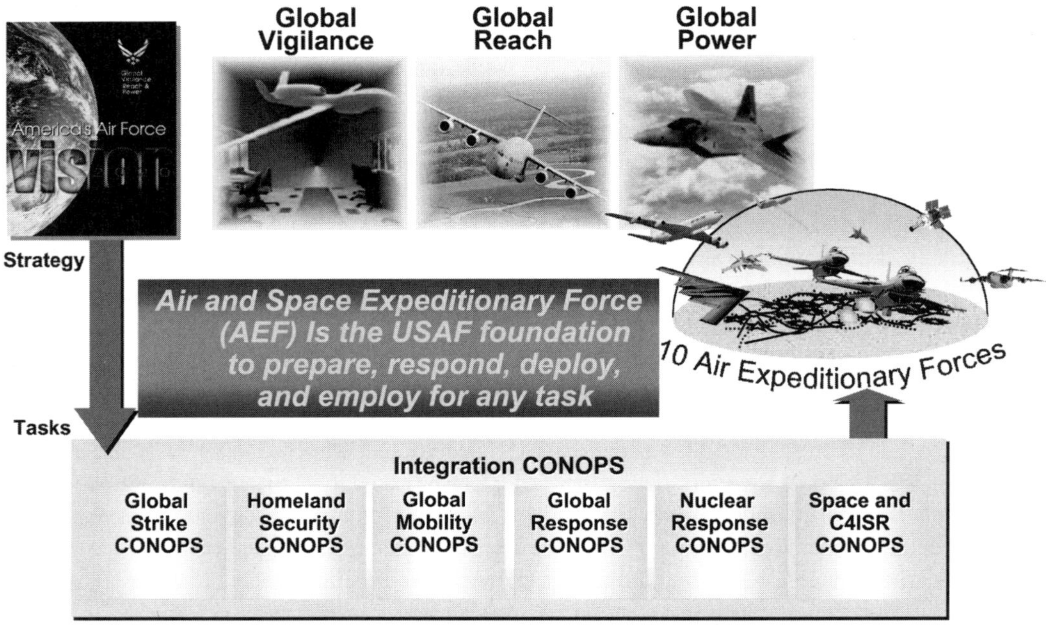

FIGURE 1-4 Air Force CONOPS construct. SOURCE: AFRL, 2003a.

space systems to provide persistent situational awareness and executable decision-quality information to the Joint Forces Commander.

Looking at the breadth of these concepts, it is easy to see how hypersonics and improved access to space can contribute to their accomplishment. These concepts also drive the programs and priorities at AFRL. A significant portion of AFRL's turbine engine technology base, for example, is focused on high-speed applications that respond directly to Air Force CONOPS-derived capability needs. Similarly, technologies derived from AFRL's Long Range Strike planning activity, with their explicit linkage to the Global Strike and Global Response CONOPS, support the development of a Mach 4+ weapon and a sustained supersonics-capable air platform, both of which are responsive to NAI's high speed/hypersonics pillar. In addition, the proposed turbine-based combined-cycle engine effort extends the technology to enable a Mach 0-4+ accelerator that addresses Air Force requirements for a responsive space launch capability, which in turn directly supports the NAI access-to-space pillar.

The AFRL rocket propulsion directorate also supports NAI and recognizes the access-to-space pillar, under which its research is being performed, as the central pillar of the program. In addition to providing rocket propulsion technology, the access-to-space pillar will provide "airframe, avionics, thermal protection system technology, operations, payload technologies and systems engineering tools" (AFRL, 2003c). It believes the technologies being pursued will be applicable and required for a myriad of vehicle configurations, including air-breathing engine/rocket or all-rocket systems, expendable or reusable, horizontal or vertical takeoff and 1.5, 2, or more stage vehicles covering the entire range of payload classes. In addition, AFRL points out that some of the technology will be applicable to payloads like the space maneuvering vehicle (SMV), the modular insertion stage (MIS), and the common aero vehicle (CAV). The bottom line is that AFRL feels strongly that work in all of these areas is important and necessary to provide an affordable, transformational access-to-space capability (AFRL, 2003c).

Operational Commands

The operational commands, including the U.S. Strategic Command, the Air Force Space Command, and the Air Combat Command, all believe that NAI may have relevance to their required operational capabilities. They believe that while the technologies and capabilities of NAI support the accomplishment of their goals and missions, other alternatives or factors should be considered. In addition, the operational commands believe that requirements to develop NAI technologies should compete with all other requirements for available resources.

U.S. Strategic Command. The U.S. Strategic Command (USSTRATCOM) believes that NAI could contribute to key components of the command's global mission of Global Strike, Global Integrated Missile Defense, Underground Facility Defeat, Mobile Missile Defeat, Global Nuclear Deterrence, Global C4ISR, and Space Operations (Shelton, 2003). USSTRATCOM believes there is potential for each of the highlighted portions of the NAI technology framework shown in Figure 1-5 to support the command's ability to accomplish its missions (Shelton, 2003).

Specifically, USSTRATCOM believes hypersonics technology could be a key enabler for its Global Strike mission because of the technology's ability to facilitate long-range rapid strike, prompt target kill, future cruise missile applications, and CAV applications. It also believes hypersonics technology could act as a deterrent and hold rogue states at risk, allow for timely penetration of enemy air defenses, and allow USSTRATCOM to launch outside the enemy's interceptor range. In addition, USSTRATCOM values the potential contribution of hypersonics technologies to its Global Integrated Missile Defense mission, because high speeds could enable the interception of incoming depressed trajectory theater ballistic missiles and cruise missiles (Shelton, 2003).

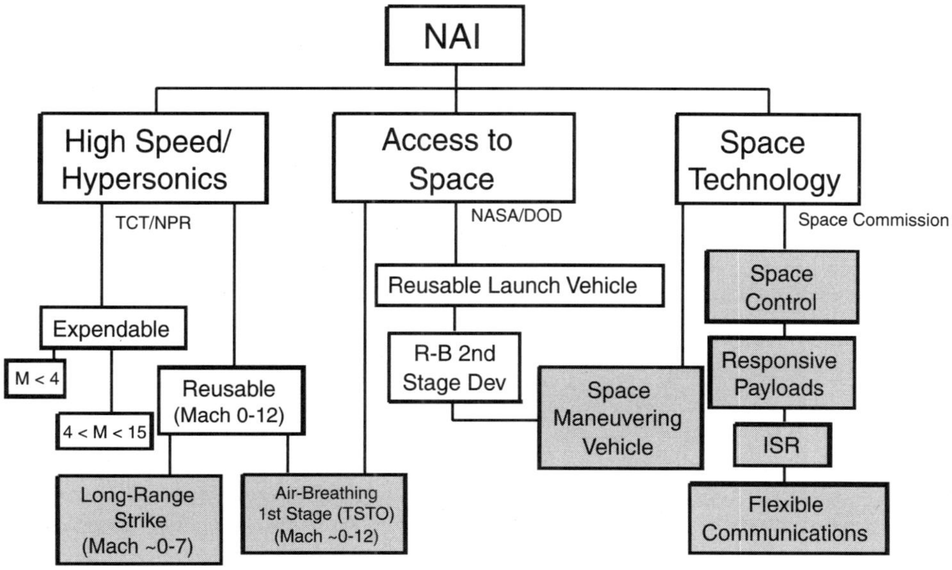

FIGURE 1-5 NAI program framework that supports USSTRATCOM missions. SOURCE: Shelton, 2003.

For the potential contributions of the NAI high speed/hypersonics pillar to be realized, however, USSTRATCOM believes that several "major technical challenges must be overcome, including the successful demonstration of hydrocarbon- and hydrogen-powered scramjet and combined-cycle propulsion concepts; the synergistic integration and control of propulsion and airframe designs; the application of material systems capable of sustaining the temperatures encountered at hypersonic speeds; and vehicle thermal management systems utilizing fuel as the principal coolant" (Winchell, 2003, p. 1).

The NAI access-to-space pillar is also of interest to USSTRATCOM because the command requires routine, on-demand access to space. USSTRATCOM believes that if the technology matures, hypersonic, air-breathing, two-stage-to-orbit capability could provide aircraftlike operations, responsive launch, and flexible launch/recovery, all at lower cost and with minimal facilities and crews. The major technical challenges that USSTRATCOM believes NAI must overcome to facilitate these contributions include "high strength, lightweight structures; advanced thermal protection systems; rocket and air-breathing propulsion; and vehicle health monitoring systems" (Winchell, 2003, p. 1).

The bottom line is that "USSTRATCOM sees potential for NAI's technological focus to support several USSTRATCOM missions. However, industry must overcome several significant technological challenges to operationalize these capabilities" (Winchell, 2003, p. 2).

Air Force Space Command. The Air Force Space Command (AFSPC) believes that NAI is relevant to the operational requirements laid out in two of its MNSs: MNS AFSPC 001-01, Operationally Responsive Spacelift (AFSPC, 2001), and MNS AFSPC 002-01, Prompt Global Strike (AFSPC, 2003). The Operationally Responsive Spacelift (ORS) MNS, approved by the Air Force Requirements Oversight Council (AFROC) in November 2001 and validated by the Joint Requirements Oversight Council (JROC) in April 2002, states that "ORS is the key enabler for conducting the full spectrum of military operations in space and for achieving space superiority" (AFSPC, 2001, p. 2). It has two subtasks:

- Transporting mission assets, including delivering payloads and changing orbits
  —Supports emerging missions: space control, missile defense, and force application and
  —Must be flexible, inexpensive, and available on demand.
- Spacecraft servicing, including traditional satellite operations such as resupply, repair, replacement, and upgrade of space assets while in orbit.

The required capabilities laid out in this MNS include the following (AFSPC, 2001, p. 2):

- On-demand satellite deployment to augment and replenish constellations,
- Launch to sustain required constellations for peacetime operations,
- Recoverable, rapid-response transport to, through, and from space, and
- Integrated space operations mission planning for on-demand execution of space operations.

The Prompt Global Strike (PGS) MNS, validated by the JROC in March 2003, has the following mission objective: "Strike globally and rapidly with joint forces against high-payoff targets in a single or multi-theater environment" (AFSPC, 2003, p. 1). According to the MNS,

> Prompt global strike (PGS) capability will give combatant commanders the capability to rapidly deny, delay, deceive, disrupt, destroy, exploit, or neutralize targets in a timeframe that is reduced from weeks and days to hours and minutes, even when U.S. and Allied forces have no permanent military presence or only limited infrastructure in a region and regardless of anti-access threats.

Prompt Global Strike mission needs include:

(1) improved responsiveness and maneuver to hold targets at risk on timelines consistent with the commander's intent and national security objectives;
(2) improved employment flexibility against preplanned and emergent targets and rapid retargeting and execution through integration of real-time intelligence updates;
(3) improved reliability and accuracy to deliver appropriate strike options to meet planned mission effectiveness criteria required by combatant commanders while minimizing collateral damage;
(4) linkage to highly accurate, complete, timely, and usable intelligence, surveillance, and reconnaissance support;
(5) survivability, to operate effectively in the defense environment and the meteorological, oceanographic, and space weather conditions that are encountered during the system's operational life;
(6) affordability, achieved by designing a system with due consideration for the life cycle costs of meeting mission requirements;
(7) robustness necessary to satisfy competing global mission requirements in a multi-theater environment. (AFSPC, 2003, p. 1)

To the extent that the technologies and capabilities emanating from NAI would support these needs, they would assist the AFSPC in accomplishing DoD-validated mission needs for ORS and PGS. Note, however, that no specific requirements for hypersonics or other NAI technologies are found in the MNS documents. Many or all of the NAI technologies might fulfill the mission needs outlined in the MNS, but the command is still looking at alternatives and has yet to determine the best ways of meeting those needs.

Air Combat Command. According to Robert Tom, for the Air Combat Command (ACC), the operational relevance of NAI is much more closely tied to the high speed/hypersonics pillar. ACC believes that high speed/hypersonics is a revolutionary capability that supports numerous CONOPS, including Global Strike, Global Persistent Attack, and, potentially, Homeland Security (Tom, 2003).

ACC, however, is focused on satisfying its near-term requirements, which do not include hypersonics. This does not diminish the command's interest in the continued development of the technology or its potential application to future long-range strike (FLRS) capabilities. ACC points out that while speed enhances survivability, it is not a stand-alone attribute: Range, payload, persistence, and survivability are also important and must be evaluated when developing a FLRS system. Potential benefits of hypersonics that interest the ACC include these (Tom, 2003, p. 1):

- Rapid response from the continental United States (CONUS) or forward operating locations;
- Increased survivability, associated with speed and altitude, against "today's" threats;
- Reduced sortie and turn time, which "may" lead to reduced fleet size;
- Ability to neutralize time-critical targets (TMTs); and
- Expanded weapons capability to include hard and deeply buried targets and weapons of mass destruction (WMD).

Air Combat Command knows that hypersonics technology is still immature and that technology is the limiting factor in fielding a hypersonics system. Other ACC concerns are the huge developmental costs and the number and type of facilities required to support the research, development, and testing of hypersonics systems. ACC believes that while the development of hypersonics technology is critical to future combat capability, it must take its place among other S&T efforts and priorities to ensure the proper balance between current, near-term, and future combat capability. Specifically, ACC feels that near-term efforts should include, but not be limited to, the following (Tom, 2003):

- Continued funding from laboratories and industry funding of hypersonic research at some level that will allow appropriate levels of research on other critical technologies,
- Continued efforts to identify critical enabling technologies that decrease hypersonics costs,
- Support for the development of a hypersonic cruise missile, as identified by the Air Armament Summit roadmap,
- Continuation of the FY 2010-2019 Hypersonic Concept Technology Development, and
- Transition to system development and demonstration by about 2020.

In summary, ACC's position is very similar to that of the other operating commands. It is interested in the technology but realizes the technical challenges ahead and the resources required. Although ACC knows that hypersonics technology is still immature, its weapons roadmap outlines a hypersonic cruise missile timeline (Tom, 2003). It believes that fielding a hypersonic weapon before a platform is the most feasible, realistic solution to its requirements. Therefore, although ACC believes that hypersonics S&T is critical for future combat capability, it also believes that the resources required to accomplish that S&T must be balanced with resources required to support other S&T efforts and priorities (Tom, 2003). Air Combat Command believes that to make an informed decision, it must evaluate all required capabilities/attributes (range, payload, speed, survivability and persistence), that trade-offs are likely, and that all solutions must compete with other ACC programs (Tom, 2003).

National Aeronautics and Space Administration

NASA's capability requirements are formally stated in PRDs that generally follow the form of standard military requirements documents. NASA's PRD NP01, Next Generation Launch Technology Program, dated August 14, 2003, lays out NASA requirements relative to NAI (NASA, 2003). All technology development and risk reduction activities previously chartered under the Second Generation Reusable Launch Vehicle Program and the Advanced Space Transportation Program were combined to form the NGLT Program. PRD NP01 defines the program-controlled requirements for the technology and systems engineering and integration projects and offices as a whole, as well as the requirements for each individual project and office under the NGLT Program:

> The objectives of NASA's NGLT Program are to identify systems to meet the national space access mission requirements and to develop technology required for national space access mission requirements. The NGLT Program is designed to advance the state-of-the-art in space transportation systems (STSs) technologies for low-cost, reliable, and safe transportation from earth to orbit; provide support to national defense needs; and support educational and public awareness efforts of the Agency. (NASA, 2003)

Specific objectives of the NGLT Program are as follows (NASA, 2003):

- Make future STSs safer and more affordable, operationally responsive, and reliable.
- Develop innovative approaches and concepts to support future decisions on systems, infrastructures, and missions for human and robotic exploration of space.
- Enhance the nation's security by developing and demonstrating critical access-to-space technologies that benefit NASA, DoD, and other government agencies.
- Improve student proficiency in space, technology, engineering, and mathematics by creating a culture of achievement using educational programs, products, and services based on NASA's unique missions, discoveries, and innovations.

• Increase public awareness and appreciation of the benefits made possible by NASA research and innovation in aerospace technology.

To achieve the objectives listed above, the NGLT program has been divided into three parts: Propulsion Technology, Launch Systems Technology, and Systems Engineering and Integration. When the objectives of NASA's NGLT Program are compared with the objectives of the National Aerospace Initiative, the commonality of interests and goals is clear.

NAI and the Interrelationship of NASA and DoD Technology Requirements

Table 1-2 shows the relationship between NASA NGLT and DoD technologies and their relevance to the NAI pillars, reusable launch vehicles (RLVs), and expendable launch vehicles (ELVs).

The overlap of various elements in the NAI access-to-space and hypersonics pillars with other major programs at DoD and NASA is evident at a glance. Although it may be difficult to find a specific requirement for each element of NAI within DoD or NASA operational requirements documents, their relevance to assigned missions and/or ongoing research programs is obvious.

Current Thrusts in NAI and Summary of End User Requirements

Currently there are three broad mission thrusts within the NAI's high speed/hypersonics and access-to-space pillars as the committee interpreted them from the stakeholders' formal statements and their supporting oral comments during presentation (Figure 1-6). Although the underlying high speed S&T base exhibits a great deal of commonality and cross-applicability among participants and their widely varying mission sets, the requirements for end application and system performance do, in fact, vary greatly, particularly between the DoD and NASA.

Hypersonics

As Figure 1-6 shows, DoD's NAI missions focus on ultra-high-speed, standoff strike and interceptor weapons against time-urgent targets, as well as on CONUS-based, global-reach, prompt, conventional strike weapons. DoD is also interested in hypersonic reconnaissance and hypersonic transport of weapons for application in-theater, but only on a wait-and-see basis. NASA missions in which global-reach hypersonic transport could play a role are driven by requirements for scheduled service economies and reliability. Importantly, not all the missions shown in Figure 1-6 receive equal attention from their primary stakeholders, nor are all parts of each service or agency equally convinced of the competitiveness of the system options being supported by the underlying technologies. For instance, neither DoD nor NASA seems much interested in ultra-high-speed (hypersonic) aircraft for the transport of either weapons or personnel over global distances. The biggest split on system concept for the same missions is between supporters of rocket approaches, which at this time are much more mature and evolutionary, and supporters of air-breathing hypersonic approaches, which need a great deal of risk-reduction before judgments can be made about them.

Space Access

DoD emphasis on space access is timely launch-on-demand and flexible basing, payload accommodation, and orbital insertion, whereas NASA emphasizes affordable recurring costs

TABLE 1-2 Relationship Between NGLT and DoD Technologies and Their Relevance to NAI Pillars, Reusable Launch Vehicles, and Expendable Launch Vehicles

Agency	Element	Common to Reusables	Hypersonics Air-Breathing Focused
NASA Augment	Air liquefaction demo		
	Hot structures	X	
	Integrated structural demo	X	
	RP engine prototype	X	
	2nd stage H_2 engine technology	X	
	Hypersonics demo follow-on		X
	High-payoff propulsion research	X	
	Integrated avionics/IVHM testbed	X	
	Power and actuation testbed	X	
NASA Inguide	Systems engineering and analysis	X	
	Vehicle systems	X	
	Integrated powerhead demo	X	
	Nontoxic OMS/RCS demo	X	
	Propulsion research and tech	X	X
	Propulsion tech and integration	X	
	X-43C vehicle (flight demo)		X
	Turboramjet Mach 4 testbed demo		
	RBCC testbed demo		X
	SSC facility	X	
	RP engine prototype demo	X	
DoD	IVHM	X	
	Streamlined operations architecture	X	
	Reusable metallic cryo tank	X	
	Durable TPS	X	
	2nd stage propulsion technology	X	
	Upper-stage engine		
	X-43C engine (flight demo)		X
	Hypersonics demo follow-on		X
	Scramjet demos		X
	Ops demonstrator	X	
	ORS CAV		

NOTE: HTHL, horizontal takeoff/horizontal landing; OSP, orbital space plane; EELV, evolved, expendable launch vehicle; ATS, automatic test system; RP, rocket propellant; IVHM, integrated vehicle health management; RBCC, rocket-based combined cycle; SSC, Stennis Space

(including logistics) and launch reliability (for crew safety and long life for reusable boost stages). NASA is willing to accept longer reaction and turnaround times than DoD.

FINDINGS AND RECOMMENDATIONS

Finding 1-1. The NAI has been very effective in fostering collaboration, communications, and advocacy of technical efforts in air-breathing hypersonics and rocket-based access to space across the services and NASA. The benefits of having the NAI are recognized by the participating organizations.

HTHL Focused	DoD Rqts Driven	Applicable to OSP	Applicable to Heavy Lift	Applicable to EELV	NAI ATS	NAI Hypersonics
X			X		X	
		X	X		X	
		X	X		X	
					X	
		X	X	X	X	
						X
			X		X	
		X	X	X	X	
		X	X	X	X	
			X	X	X	
		X	X	X	X	
			X		X	
		X	X		X	
			X		X	
			X	X	X	
						X
X			X			X
						X
			X	X	X	
		X	X	X	X	
	X	X	X	X	X	
	X	X	X	X	X	
			X		X	
	X	X	X		X	
	X	X	X	X	X	
	X		X	X	X	
						X
						X
						X
	X		X	X	X	
	X				X	

Center; TPS, thermal protection system; ORS, Operationally Responsive Spacelift; CAV, common aero vehicle.
SOURCE: Lyles, 2003.

Recommendation 1-1. DoD and NASA, employing NAI, should continue to advocate, communicate, and facilitate the nation's endeavors in hypersonics and access to space and encourage a global view beyond the institutional constraints imposed on individual partners.

Finding 1-2. Incomplete data limit the ability to obtain a clear, comprehensive picture of NAI funding, both by those who would assess the initiative's financial feasibility in a specified time frame and by NAI's participants. The lack of comprehensive funding data complicates effective, coordinated use of resources across the NAI's many programs and activities.

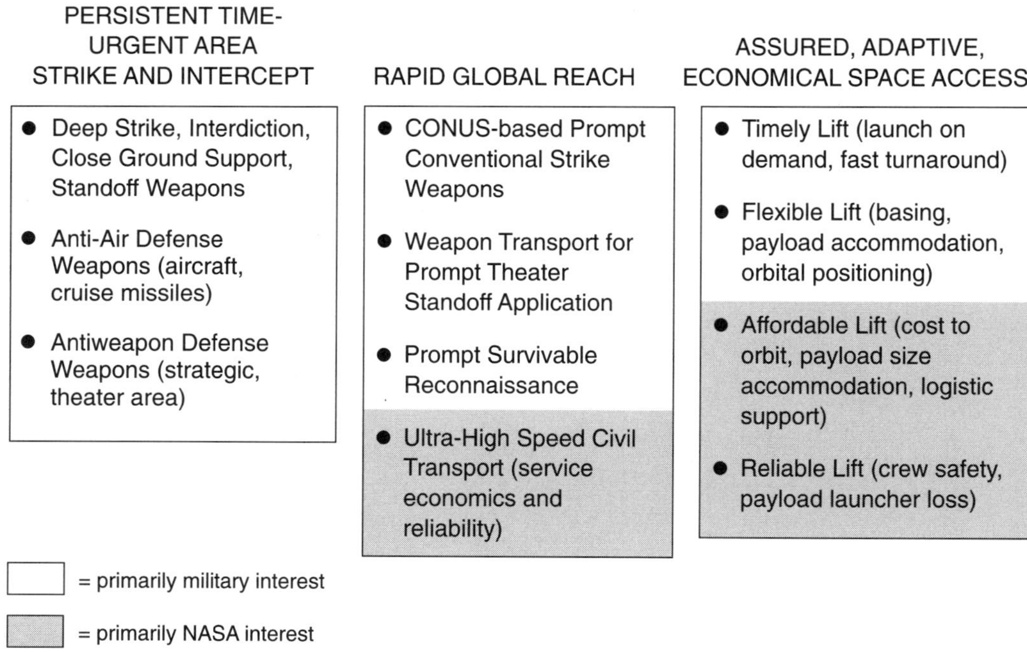

FIGURE 1-6 NAI missions for NAI hypersonics and space access.

Recommendation 1-2. The NAI executive office should develop and communicate clear, comprehensive NAI funding data that show all the ongoing, planned, and newly proposed NAI projects of DoD and NASA depicted on NAI roadmaps and tied directly to objectives. In their budget plans and submissions, DoD and NASA should show all NAI projects and budget lines as part of a unified NAI effort.

Finding 1-3. The military services and NASA are generally supportive of the technologies and capabilities envisioned by NAI. They see those technologies and capabilities, if they mature, as being relevant to the capability requirements they have developed in support of their assigned mission requirements. They believe these initiatives need to be supported and funded in proportion to their potential contribution to the current and future operational requirements of each organization. The variance in those requirements, resource limitations, unknowns surrounding R&D results and timing, and different assessments of the applicability of the resulting technologies are all part of the stakeholders' current uncertainty with respect to NAI.

Finding 1-4. The services and NASA believe NAI's efforts to sustain our nation's aerospace leadership through technology development and demonstration are of great value in terms of focusing research, advocating funding, facilitating coordination, and keeping the lines of communication open to ensure that informed decisions are made. However, the services expressed concern about any NAI authority to direct or coordinate the research itself, because NAI is not in a position to evaluate internal competing service and NASA requirements. NAI can play a strong advocacy role by fostering research that will ensure our country's continued aerospace leadership and facilitating the efficient accomplishment of that research.

REFERENCES

Published

AFSPC (Air Force Space Command). 2001. FINAL Mission Needs Statement AFSPC 001-01, Operationally Responsive Spacelift, ACAT Level I. Colorado Springs, Colo.: AFSPC.

AFSPC. 2003. FINAL Mission Need Statement AFSPC 002-01, Prompt Global Strike, ACAT I. Colorado Springs, Colo.: AFSPC.

DDR&E (Director of Defense Research and Engineering). 2003. National Aerospace Initiative, Sustaining American Aerospace Initiative. Available at http://www.dod.mil/ddre/nai/nai_brochure1.pdf. Accessed on December 23, 2003.

NASA (National Aeronautics and Space Administration). 2003. Program Requirements Document for the Next Generation Launch Technology Program. NP01, Version Revision A. NGLT-PROG-0001A.

Unpublished

AFRL (Air Force Research Laboratory). 2003a. Response to committee questions at Dayton meeting. Sent by Jack Vondrell to Lt. Gen. Kenneth Eickmann, committee member. September 23.

AFRL. 2003b. Operational Relevance of NAI Responsive Turbine Engine Technologies. Prepared by Tim Lewis and sent by Jack Vondrell to Lt. Gen. Kenneth Eickmann, committee member. September 23.

AFRL. 2003c. NAI Rocket Propulsion Point Paper for Air Force Science and Technology Board. Sent by Jack Vondrell to Lt. Gen. Kenneth Eickmann, committee member. September 18.

Blackhurst, J. 2003. AFRL and NAI Perspectives to AFSTB Committee on NAI. Briefing by Col. Blackhurst, Chief, Space and Missile Systems Sector, AFRL, to the Committee on the National Aerospace Initiative. September 3.

Liston, G.W., C.R. McClinton, M.L. Mumford. 2003. High Speed/Hypersonics Pillar Technology Development Approach. Presentation by Glenn Liston, NAI Executive Office, to the Committee on the National Aerospace Initiative. October 23.

Lyles, G. 2003. FY 04-09 Application Matrix. Prepared by Garry Lyles, NGLT Program Manager, NASA, in response to questions by the committee. Submitted on October 8.

Richman, M. 2003a. National Aerospace Initiative: Sustaining American Aerospace Leadership. Briefing by Mike Richman, Special Assistant to the DDR&E, to the Air Force Science and Technology Board. March 24.

Richman, M. 2003b. National Aerospace Initiative. Briefing by Mike Richman, Special Assistant to the DDR&E, to the Committee on the National Aerospace Initiative. October 7.

Rogacki, J. 2003. The National Aerospace Initiative—A Synergistic Partnership Between NASA and DoD. Briefing by John "Row" Rogacki, Deputy Associate Administrator for Space Transportation, Office of Aerospace Technology, NASA, to the Committee on the National Aerospace Initiative. August 5.

Sega, R. 2003. National Aerospace Initiative—NRC Briefing. Briefing by Ron Sega, DDR&E, Office of the Secretary of Defense, to the Committee on the National Aerospace Initiative. August 6.

Shelton, W.L. 2003. National Aerospace Initiative (NAI) USSTRATCOM Perspective. Briefing by Brig. Gen. Shelton, Director, Policy Resources and Requirements, USSTRATCOM, to the Committee on the National Aerospace Initiative. September 3.

Tom, R.C. 2003. ACC Bullet Background Paper on Speed Versus Other Future Long Range Strike (FLRS) Attributes Needed by the Warfighter. Prepared by Robert C. Tom, ACC/DRMA(ANSER)/4-5832/ret/6 Oct 03, and given to Lt. Gen. Kenneth Eickmann, committee member. October 6.

Winchell, G. 2003. National Aerospace Initiative (NAI) Information Paper. Paper prepared by Maj. Winchell, USAF, for Lt. Gen. Kenneth Eickmann, committee member. October 7.

2

Hypersonic Flight

INTRODUCTION

Air-breathing hypersonic flight has long been recognized for its potential to enable both advanced military missions and (possibly) efficient access to space. It offers two principal advantages: high speeds, with associated reduction in time to target, and the potential for substantial performance gains resulting from the high specific impulse (Isp) achievable with air-breathing propulsion. However, these potential advantages must be weighed against formidable practical challenges, such as those posed by internal and external aerodynamics, supersonic mixing and combustion, and the costs of development as well as the requirements for high-temperature materials, lightweight airframes, and stability and control. Over the past 40 years, a long series of development programs has been funded in the United States and in other countries. Although these programs have generated a great deal of theoretical and practical understanding, sustained air-breathing propulsion and flight at high Mach numbers has remained elusive. Clearly, much more work needs to be done along many fronts.

On the other hand, scramjet technology has matured significantly over the past decade or so, and the flight-testing of scramjet engines is imminent. Computational methods such as computational fluid dynamics (CFD), finite element model (FEM) structural/thermal analysis, and vehicle design/optimization methods have also matured significantly, making it more likely that highly integrated hypersonics systems can be designed to close with respect to performance/economic requirements. In addition, command, control, communications, computing, intelligence, surveillance and reconnaissance (C4ISR) technologies have developed to the point where the timeline for finding, selecting, and engaging targets (the "kill chain") is getting short enough to take advantage of the rapid response afforded by hypersonic missiles and reentry vehicles.

Recognizing the potentially large benefits of air-breathing hypersonics technology and the magnitude of the efforts to develop this technology, the NAI brought together essentially all existing U.S. programs and attempted to lay out a high speed/hypersonics technology roadmap. However, this roadmap is more a collection of existing programs than a logical and complete plan for achieving military strike, global reach, and space access objectives. Each of the collected programs has elements that address some portion of the critical technologies, and each has a funding line. The collective funds, if properly applied, may be sufficient in the near term for a significant critical technologies program; however, as configured, it is not clear that the collected programs cover all critical hypersonics technologies. What is more, NAI has not projected funding after existing program budgets run out. The committee observes that sharply higher budgets will be required to support technology scale-up (i.e., beyond presently planned small-scale demonstrations), especially flight demonstration programs aimed at maturing air-breathing hypersonics technologies to the point where decisions can be made about the development of large-scale Global Strike/ISR aircraft and air-breathing space access vehicles. A realistic budget must be projected consistent with answering the critical technology challenges, and existing programs should be more closely aligned, taking advantage of the synergistic potential championed by NAI, so they can provide critical hypersonics technologies to the nation.

This chapter begins by presenting the committee's findings and recommendations, with brief discussions as appropriate. Technical and financial issues associated with the technologies that are critical to hypersonic flight are then discussed, together with recommended directions for future research and development.

FINDINGS AND RECOMMENDATIONS

Finding 2-1. The U.S. Air Force, the U.S. Army, and the U.S. Navy all see the possible benefits of applying air-breathing hypersonic propulsion technology to a broad range of warfighting missions, but none has yet developed formal requirements for such technology. Similarly, NASA sees potential in applying air-breathing hypersonics propulsion technology to space launch systems. Together, the DoD services and NASA are investing in the development of near- and mid-term hypersonics technologies under the NAI and are looking to 2018[1] as a point at which to assess whether hypersonic propulsion has sufficient system and/or operational benefits to warrant applying the technology to space launch missions (Sega, 2003a). The DoD services see air-breathing hypersonics technology as applied to hypersonic missiles and/or aircraft as tangible products along the path to 2018 (see Chapter 1 for more discussion of warfighter requirements).

Discussion 2-1. Although they recognize the potential benefits of applying hypersonic air-breathing technology to weapon systems, the DoD services are expressing a wait-and-see attitude, keeping an eye on hypersonics technology development, and continuing to explore concept of operations (CONOPS) for hypersonics systems employing air-breathing propulsion (Morrish, 2003; Graff, 2003; Hickman et al., 2003; Walker, 2003). For example, the Air Force recognizes that its space access requirements (as well as its time-critical strike and global reach requirements) may someday be met by hypersonic air-breathing propulsion, but only after the technology has been sufficiently matured. In the meantime, the greater maturity of rocket technology allows it to develop near- and mid-term (~2010 and 2015, respectively) rocket-based solutions to satisfy Operationally Responsive Spacelift (ORS) requirements. By 2018, hypersonic air-breathing technology may be sufficiently mature and understood, under a properly developed and executed NAI plan, to make a full-scale development (FSD) decision on whether to promote air-breathing hypersonics technology to the next block or spiral of space access system development, one with an initial operational capability (IOC) no sooner than about 2025.

---

[1]This date and others that appear in the report were established by NAI participants prior to the President's announcement of a new mandate for NASA. How these dates will be affected by the new NASA mandate is not yet clear; however, some NAI schedule objectives might be significantly delayed. See "NASA's New Space Exploration Mandate" in the preface.

NASA is also aiming for an FSD decision in 2018 on applying hypersonic air-breathing propulsion to space launch systems (Lyles, 2003; Rogacki, 2003), but it views the technology as a potential contributor to its space access future; on the other hand, the Air Force views the technology as just one of several options that may or may not get applied to far-term launch vehicles.

To support near- and mid-term air-breathing hypersonic applications and an FSD decision in 2018 on whether to employ hypersonic propulsion on a subsequent generation of space access systems, the DoD and NASA would have to complete the planning, funding and execution of the NAI. All requisite technologies and an associated plan must be developed sufficiently to permit hypersonic product development decisions to be made based upon real data.

Recommendation 2-1. The DoD should continue to evaluate the unique capabilities achievable with air-breathing hypersonic systems as a means of satisfying warfighting needs. Despite the current lack of formal requirements, the DoD and NASA should also continue to invest in the development of air-breathing hypersonic technologies to meet their future capability needs.

Finding 2-2. The objectives of the high speed/hypersonics (HS/H) pillar of the NAI are to develop near- and mid-term hypersonic technologies to the point where they could support emerging DoD mission capability requirements and, far term, to evolve an air-breathing, hypersonic, reusable space launch system for both NASA and DoD applications. However, the top-level HS/H plan (roadmap) for accomplishing these objectives does not appear to be coherent, comprehensive, or well communicated to decision makers and stakeholders.

Discussion 2-2. The NAI program claims to rely heavily on a formalized decision process employing goals, objectives, technical challenges, and approaches (GOTChA) to establish the content of HS/H technology development and flight demonstration roadmaps in support of an air-breathing launch vehicle FSD decision in 2018 (see Figure 2-1). However, many elements of the NAI program were in place prior to GOTChA, and the process did not appear to change these preexisting elements. It is not clear that the process, as used by NAI, was effective in defining a comprehensive and compelling technology development program, from fundamental research to flight demonstration, that would sufficiently mature all technologies critical to operational hypersonic flight. Particularly noteworthy is the low level of fundamental research (6.1 and 6.2) in the NAI HS/H plan. Although two NASA university research, engineering, and technology institutes (URETIs) have been established to sponsor fundamental research in air-breathing launch vehicle technologies, this level of support is deemed insufficient to support NAI goals. Similarly, it is not clear that comprehensive plans yet exist to develop enabling technologies and critical system components to the point where operational hypersonic flight can be achieved in the time frame specified. And finally, recognizing the need to conduct flight demonstrations to advance hypersonics technology, it is not apparent that NAI has identified the set of flight demonstrations necessary and sufficient to support an FSD decision in 2018.

In summary, the HS/H program appears to be primarily a collection of preexisting flight demonstration programs or concepts slated to be used to mature hypersonics technologies. Regardless, the committee believes the NAI hypersonics plan to be potentially technically feasible in the time frame laid out, but only if the necessary planning occurs, adequate funding is available, the beneficial effects of synergy are successfully captured by the NAI partners, and science enables breakthroughs in the associated critical technologies.

Recommendation 2-2. Starting with a defined and articulated vision, DoD and NASA should use a top-down process based on sound system engineering principles to determine the objectives, technical challenges, and enabling technologies, and the fundamental research, technology devel-

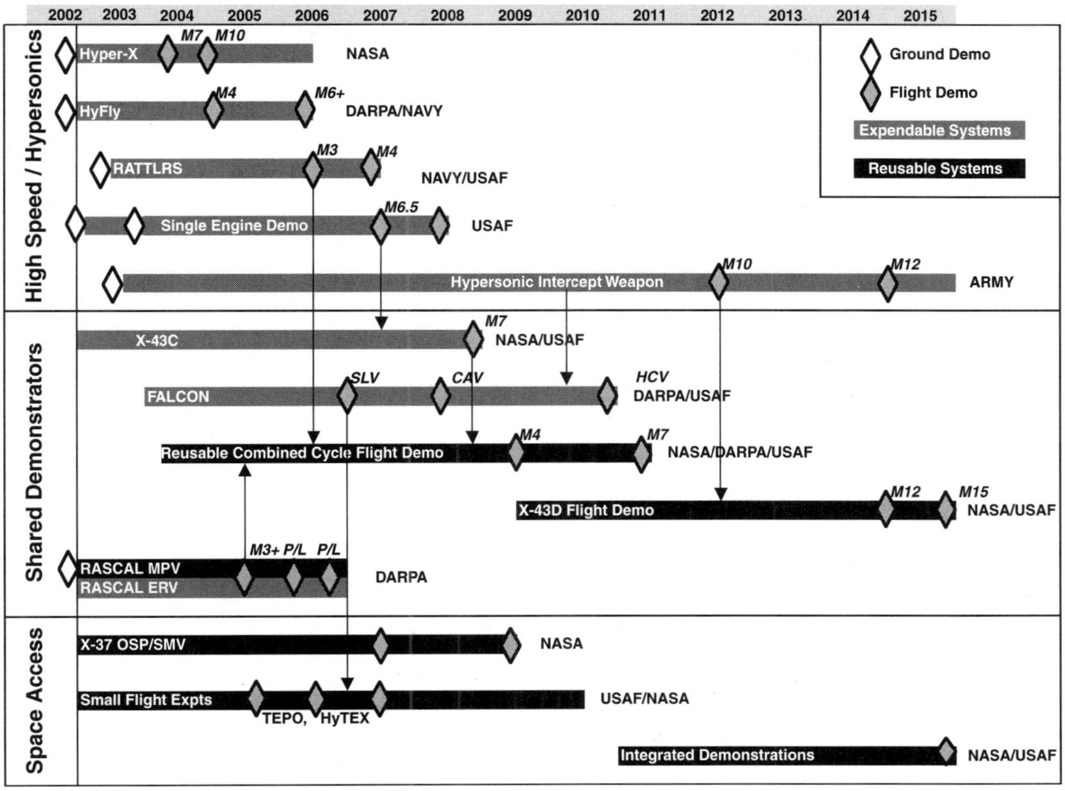

FIGURE 2-1 NAI integrated high speed/hypersonics and access-to-space ground and flight demonstration roadmap. SOURCE: Sega, 2003b.

opment, ground testing, and flight demonstration plans required to mature enabling technologies to the point where they can be applied to operational hypersonic flight. The result should be a comprehensive integrated roadmap that assures sufficient maturation of all critical technologies prior to making FSD decisions on space access in 2018 and prior to proceeding with earlier possible hypersonics applications as well (e.g., missiles and aircraft).

The end product should be a roadmap similar to the notional one presented in Figure 2-2 but also including detailed roadmaps for fundamental research and each of the critical technologies. Once complete, the plan should be clearly communicated to decision makers and stakeholders, including the public. Unless this detailed planning is done, insufficient information will exist upon which to base sound implementation and funding decisions.

Finding 2-3. Based on information provided to the committee, the current level and stability of NAI funding for air-breathing hypersonics S&T are insufficient to achieve the NAI goals by 2018.

Discussion 2-3. Using available data, the committee identified four critical and enabling technologies for air-breathing hypersonic flight that must be sufficiently matured to a technology readiness level (TRL) of 6 or 7 to support the near-term development of missile systems, the medium-term development of aircraft systems, and an FSD decision on access to space in 2018 (see Box 2-1).

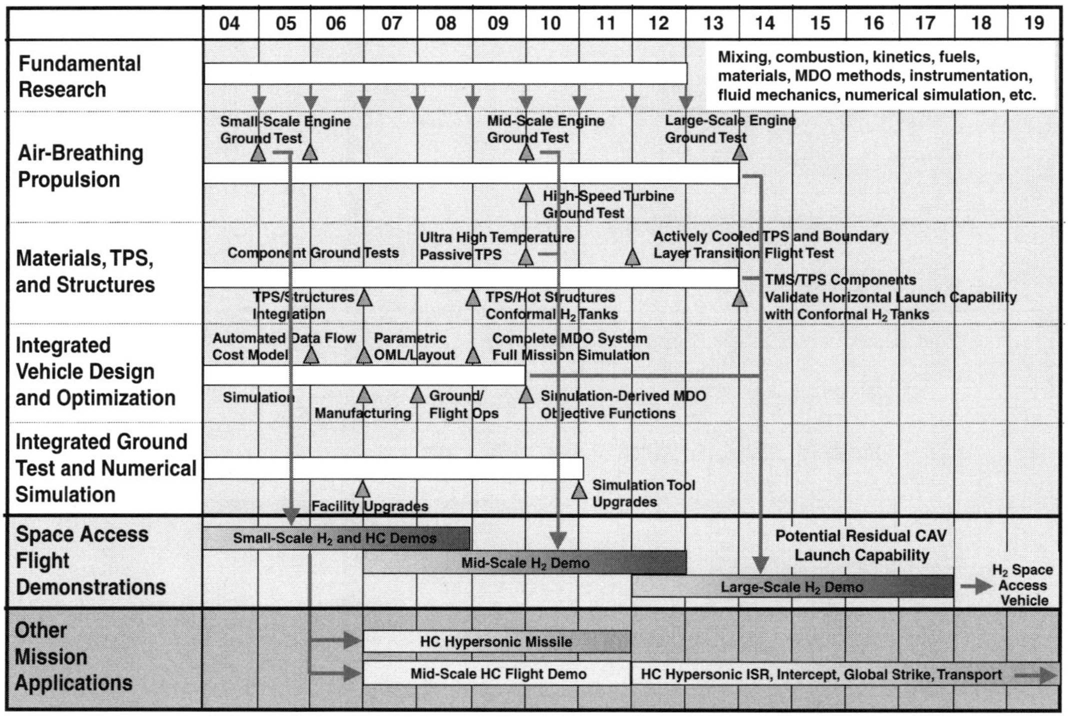

FIGURE 2-2 Notional comprehensive hypersonic flight roadmap showing all requisite activities, from fundamental research to critical technology development to flight demonstration, for making a decision in 2018 on full-scale development of hypersonic air-breathing access to space systems. SOURCE: Adapted from Bowcutt, 2003.

However, in the committee's judgment, there is insufficient funding for maturing these technologies. The four critical technologies are these:

- Air-breathing propulsion and flight test,
- Materials, thermal protection systems (TPSs), and structures,
- Integrated vehicle design and multidisciplinary optimization, and
- Integrated ground testing and numerical simulation/analysis.

A detailed discussion of each of these technologies, including assessments of current TRLs, technical feasibility in the NAI time frame, available and required budgets, and recommendations for near- and far-term technical emphasis, can be found in the next section, "Hypersonic Flight Critical Technologies."

Recommendation 2-3. DoD and NASA should complete an end-to-end cost estimate for the top-down program through the period of interest and then work to establish funding commitments consistent with the plan.

Finding 2-4. The NAI has been very effective in fostering collaboration, communication, and advocacy of detailed technology development in air-breathing hypersonics across the services and

Box 2-1 Technology Readiness Levels

Technology readiness levels (TRLs) refer to a system of metrics that is used to assess the maturity of a technology, compare the maturity of different technologies, and communicate technology maturity to others. The NASA TRL scale, which was developed for flight hardware items, can be seen at http://spacescience.nasa.gov/admin/pubs/handbook/OSSHandbook.pdf (accessed January 2, 2004).

The committee used an extension of the standard NASA TRL scale to develop consistent TRL ratings for all the technologies reviewed, nonhardware items as well as hardware items.

	TRL	Product	Process	Analysis/Simulation
Implementation	9	Actual System "Flight Proven" Through Successful Mission Ops.	Actual Process Proven Through Successful Operation by Program	Actual Models in Use by the Community
Validation/ Verification	8	Actual System "Flight Qualified" Through Test & Demo	Actual Process Completed and "Qualified" Through Test/Demo	Actual Models Validated Against "Flight Qualified" Data
	7	System Prototype Demonstration in an Operating Environment	Prototype Process Demo in a Program Environment	Prototype Model Validated Against Flight-Test Data
Demonstration	6	System/ Subsystem Prototype Demo in a Relevant Environment	Process Prototype Demo in a Relevant Environment	Model Validated Against Relevant Ground-Test Data
	5	Component Validation in Relevant Environment	Beta Version Key Elements Validated In Relevant Env.	Model Components Evaluated Against Relevant Data
Development	4	Component Validation in Laboratory Environment	Alpha Version Key Elements Validated Against Benchmark	Tools Assembled into Package and Tested Against Hand Calcs.
Proving Feasibility	3	Critical Function of Characteristic Proof-of-Concept	Alpha Version Operational in a Test Environment	Data Flow Diagrams, Tools Collection and Familiarization
Basic Research	2	Technology Concept and/or Application Formulated	Requirements Document Approved by Customer	Methods and Algorithms for Similar Systems Identified
	1	Basic Principles Observed and Reported	Current Process Documents and Potential Savings Identified	System Characterized and Tool Needs Defined

SOURCE: Bowcutt, 2003.

NASA. While this has great potential for leveraging existing funding to the benefit of all stakeholders, there may be duplication in some of the contributing NAI programs.

Discussion 2-4. The services, the Office of the Secretary of Defense (OSD), and NASA are highly engaged in formulating the NAI program. This has helped to tighten the coupling among the S&T communities within the services and NASA and encouraged the communication and collaboration that is now leading to a more efficient and rationalized NAI. This collaboration includes the warfighting community, although the formal process for formulating service requirements is not necessarily suited to the cross-agency focus of NAI. (See the discussion of warfighter requirements in Chapter 1.)

Examples of cross-agency collaboration include the Air Force's Hypersonics Technology (HyTech) and Single-Engine Demonstrator (SED) programs and NASA's X-43C program. With the HyTech program, the Air Force is developing and ground testing a hydrocarbon scramjet engine. The Air Force SED program plans to flight test a single fixed-geometry HyTech engine module in an expendable demonstration vehicle. NASA's X-43C program plans to flight test three variable-geometry HyTech engine modules, also in an expendable demonstration vehicle.

The goals and objectives of the X-43C and SED programs, as planned at the time this study was conducted, were similar. The most important technology demonstration for both the X-43C and SED programs is engine operability and performance across the operating range of the scramjet: from ignition, through ramjet-scramjet transition, to acceleration and cruise. Both programs share the same engine core—namely, the hydrocarbon-fueled scramjet of the Air Force's HyTech program. A HyTech engine flight demonstration is particularly critical because of the use of hydrocarbon fuel and the need to use the endothermic capacity of the fuel to maintain heat balance during cruising flight (see discussion of hydrocarbon fuels in Discussion 2-7 and Appendix E). Other shared goals of the two programs are airframe, engine, and system integration; thermal protection system approaches; component durability; thrust, drag, and efficiency predictions; and flight control and management.

The committee was impressed not only with the progress of the engine development program but also with the management structure and government participation in the program. In the committee's opinion, the partnership between the government program managers and the contractors provides an excellent example of how such a management model can lead to rapid and continuous technology improvement.

As far as this committee was able to determine, the primary difference between the planned X-43C and SED programs is the demonstration of multiple engine flow paths versus a single engine. The X-43C multiple-engine design addresses two propulsion technology issues not shared by the SED program: the simultaneous control of multiple engines and a variable-geometry inlet required to effect engine control.

Under NAI, it might be possible for the Air Force and NASA to collaborate even further on an expendable vehicle flight demonstration. Instead of separate flight tests, the Air Force and NASA might be able to devise an alternative approach that would achieve the combined goals and objectives of the two similar programs. For example, development and incorporation of an engine control system that prevents engine flameout or inlet unstart and demonstration on a single-engine module in flight might mitigate the risk of multiengine operation sufficiently to permit the use of multiple engines in a later reusable flight demonstration.

The committee notes that canceling or restructuring existing programs can be politically and contractually sensitive. The potential benefits of doing so must be weighed against the real-world costs. There can be strong, legitimate reasons for conducting development and test programs with similar objectives.

Recommendation 2-4. NAI should continue in its role of advocating, communicating, and facilitating the elements of the nation's HS/H and space access endeavors and should encourage a global view beyond the institutional constraints imposed on the individual partners. In this role, NAI should also identify and eliminate unnecessary duplication of planned activities to maximize the utilization of resources available for achieving NAI goals. Specifically, NAI programs should be critically reviewed to discern if closer collaboration and consolidation might eliminate potentially wasteful duplication.

Finding 2-5. The high heat loads of air-breathing hypersonic flight place severe demands on the materials of which the vehicle, its propulsion system, and the mission sensors are made. Although significant advances have been made in materials for both active and passive thermal protection systems, much work remains to be done. The TRL assessments for materials presented to the committee by the Air Force and NASA were markedly different.

Discussion 2-5. Hypersonic flight poses well-recognized challenges in terms of materials that will survive the high temperatures and high heat loads imposed by flight environments. For example, at speeds above about Mach 8 (and at lower speeds inside the engine), some regions on hypersonic vehicles will exceed current material temperature limits of approximately 3000°F, making active cooling of even the most heat-resistant materials a requirement in these areas.

Despite the challenges, much progress in materials development has been made. During the National Aerospace Plane (NASP) program, a dedicated NASP materials and structures augmentation program was funded as a part of the overall effort (SAIC, 1995). The extensive development and testing from this program have significantly extended the range of materials available for application to hypersonic vehicles, but much work remains to be done to fully qualify the wide range of materials required for all the anticipated needs of vehicles and their propulsion systems. Specific materials issues are discussed at greater length in the subsection on materials.

NASA has developed a structured approach to classifying the TRL for various technologies including materials for hypersonics systems. This appears to be an excellent way to ensure that an objective cross-technology approach is used to judge the development of a given technology. In a presentation to the committee, a representative of NASA Langley (McClinton, 2003) provided data indicating that the TRLs for many of the materials required for a reusable Mach 7 hypersonic demonstrator vehicle are between 4 and 6.

On the other hand, in a presentation made on the same day to the committee, a representative of AFRL (Evans, 2003) indicated that the technology readiness of the materials for the same vehicle and propulsion system was much lower, though the rating system used by the Air Force was not numerical and therefore difficult to correlate. Both NASA and the Air Force made the point that significant effort will be needed to develop materials for high heat flux applications such as leading edges and combustor panels. However, the committee tended to agree more with the Air Force assessment—that is, that the materials currently available are not sufficiently mature to ensure that a successful program can be executed with an acceptable level of risk.[2]

In spite of this situation, significant advances have been made in certain critical materials technology areas. Some of these advances were not presented in any detail to the committee, but they should be mentioned here as they could have great impact on the success of future hypersonics systems. Two examples of such technologies are toughened ceramic tiles and actively cooled carbon-silicon carbide ceramic composites. Both are recent advances that hold significant promise and will be discussed at greater length in the materials subsection of this chapter.

Another important aspect of materials for hypersonic systems is their supply chain. Many of the specialized materials that have been studied for such applications have only been produced in limited quantities. Examples of such materials are high-conductivity composites, titanium matrix composites, and actively cooled ceramic composites. If these materials are to be used in large quantities, qualified sources of high-quality material will have to be developed.

Recommendation 2-5. Aggressive materials technology development should continue at all participating NAI agencies, in line with NAI goals and objectives. All agencies and contractors working on technologies applicable to hypersonics, and in particular those working on the development and application of materials, should uniformly adopt the NASA TRL system of assessing technology readiness and reconcile differences in technology evaluations.

---

[2]The committee discussed the reasons for the different perspectives of the Air Force and NASA and concluded that one reason was a difference in requirements and the other was a matter of communications. Suggestions for addressing the communication issue are proposed in Recommendation 2-8.

Finding 2-6. The NAI plan does not appear to contain sufficient activity in combined ground testing and numerical simulation for hypersonics technology development.

Discussion 2-6. Only a few hypersonics high-enthalpy facilities exist in the United States. They all suffer from limitations. Facilities used to test scramjet engines for relatively long duration are vitiated (impure) air tunnels in which the free stream contains combustion products and in which the enthalpy is limited to below Mach 8. The shock tunnels and expansion tunnel at the Calspan-University of Buffalo Research Center, Inc. (CUBRC), the Caltech T5 shock tunnel, and the General Applied Sciences Laboratory (GASL) expansion tube range in enthalpy up to Mach 20 but are all short-duration facilities (1 to 10 ms) and also have other limitations, including free stream dissociation. Nevertheless, through a balanced combination of hypersonics high-enthalpy facility testing and numerical simulation, tunnel free stream conditions can be well characterized, and new and important effects can be discovered that will contribute to the development of vehicle design tools.

New flow variables that come with high enthalpy, such as vibrational excitation and species concentration, can be measured by modern optical diagnostic techniques, which have, however, only been applied to high-enthalpy ground testing facilities in a few very limited cases. More detailed measurements in the free stream and in the flow fields of tested articles—particularly in the engine combustor of direct-connect tests—could provide essential data for validating simulation methods. A further technology shortfall is an inadequate knowledge of reaction rates, in particular the coupling of vibrational excitation, dissociation, and surface chemistry.

The large number of dimensionless variables involved in high-enthalpy flows makes it impractical to develop a vehicle by testing in cold hypersonic and visciated air facilities alone. Ground testing in hypersonics high-enthalpy facilities using modern optical diagnostic methods, closely coupled with computational investigations of the same flow, is needed to develop numerical simulation tools that take proper account of high-enthalpy effects. Such tools can then be used with greater confidence in the design and preparation of flight tests.

Recommendation 2-6. NAI's plan should include as a substantial part of the HS/H pillar the development of combined ground testing and numerical simulation technologies which provide better and more reliable design tools for high-enthalpy flows. Such technology development requires much more 6.1 funding than is now available.

Finding 2-7. In hydrocarbon-fueled hypersonic engines, the endothermicity of the fuel plays a critical role in cooling the system, limiting the maximum operational speed of systems using such engines. In contrast to this challenge are the operational benefits of storable fuels such as hydrocarbons, yet we find that no continuing fuels development program is being funded in the NAI that might increase the endothermic capacity of storable fuels and permit higher operational speeds.

Discussion 2-7. Although not critical to the achievement of NAI's present Mach number objectives for hydrocarbon-based, air-breathing hypersonic propulsion, a comprehensive research and development program in endothermic fuels could remove the Mach 7 upper speed limit imposed by relying on JP-7 fuel. Higher maximum speeds using hydrocarbon fuels might enable the application of such fuels for access-to-space missions while retaining their inherent logistical advantages.

The maximum Mach number of a hypersonic cruise vehicle is determined by its heat-balance capability at sustained cruise speed. The heat has two sources: the recovery temperatures present in boundary layers over the vehicle and in its inlet, and the conversion of the fuel's chemical energy to thermal energy via the combustion process inside the scramjet engine and in some cases extending into the exit nozzle. Clearly, flow path designs can play an important role in minimizing the heat load by, for example, minimizing surface-to-flow area configurations. Low-heat-rejection materials and structures must be pursued for the combustor so that it can run hotter, significantly reducing the heat removal requirement over the present 2000°F obtainable using superalloys (cf. subsection on materials). In the end, only the fuel itself (and to a minor extent radiation) is available for cooling. More information on fuels and a possible fuels development program can be found in Appendix E.

Recommendation 2-7. A long-term and sustained fuels program should be established to explore the Mach number limits of hydrocarbon and other storable fuels.

Finding 2-8. No independent advisory board has been formed to help NAI establish and achieve program goals, objectives, and planning, as called for in the documentation provided to the committee (Sega, 2003a) and shown in Figure 2-3.

Discussion 2-8. The committee was generally impressed with the progress that the NAI has made in facilitating the coordination of the participant's activities. However, it believed that this coordination effort would benefit from periodic oversight of NAI by an independent group of experts.

Recommendation 2-8. DoD and NASA should set up their planned advisory panels, steering groups, and revolutionary concepts panels to review the NAI program on a continuing basis. Related to this, NAI should form a critical technology coordination office for each of the four enabling hypersonic technologies, the purpose of which would be to enhance coordination of development efforts among the different agencies and contractors.

Finding 2-9. Survival of a long-term initiative like the NAI depends on tangible technical products being produced along the way, yet many of the potentially viable weapons and other applications

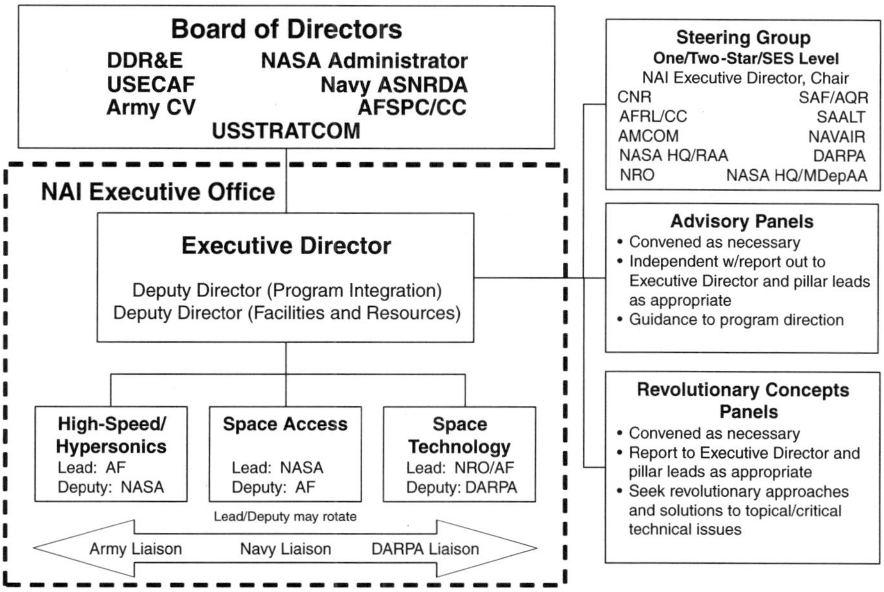

FIGURE 2-3 Structure of the NAI Executive Director's Office. SOURCE: Sega, 2003a.

enabled by air-breathing hypersonics technologies, which are clearly available along the roadmap to space access, are not sufficiently (or equally) emphasized in the stated goals of the program.

Discussion 2-9. One of the committee's most difficult tasks was trying to concisely state the objective of the NAI. To sustain a national commitment to such a long-term goal, there must be a perception that the nation is continuing to benefit at each step of the way. Essentially all of the roadmaps reviewed by this committee indicated that there are many technologies in the early stages of development that could be matured and applied to DoD near- and mid-term products that address quick response and global reach. An example is a missile that could be developed based on the HyFly or SED demonstrator vehicles, both of which will fly in the next 2 or 3 years. Supporting this possibility, C4ISR, a critical enabler for the effective use of hypersonic missiles, has developed to the point where the timeline for finding, selecting, and engaging targets (the "kill chain") is becoming short enough to take advantage of the rapid response afforded by hypersonic missiles. Despite the possibilities, these early spin-off products were not emphasized very strongly as being specifically part of the goals and objectives of NAI. Rather, they were discussed in a casual, offhand fashion.

Recommendation 2-9. NAI should develop a concise description of its goals and objectives, giving equal weight to early applications achievable in the near and medium term, including hypersonic missiles and aircraft that are already part of the HS/H pillar.

Recommendations 2-10, 2-11, and 2-12 are found in the first subsection of the section that follows, "Hypersonic Flight Critical Technologies."

HYPERSONIC FLIGHT CRITICAL TECHNOLOGIES

This section covers the technologies that are critical for air-breathing hypersonic flight and discusses their current readiness levels. Because some of these technologies are analysis tools, processes, or ground tests—that is, not directly part of a flight system—the standard NASA TRL scale was extended to develop consistent TRL ratings for all technologies reviewed (see Box 2-1). The committee notes, however, that achieving high TRLs in the various technologies associated with air-breathing hypersonics does not guarantee that the system-level goals will be achieved. For example, one might be able to cool surfaces exposed to air with recovery temperatures of 6000 K, but will one be able to turn around a vehicle within 12 hours of such a flight? Can a reusable vehicle be refurbished for $1 million dollars or less? High TRLs help, but they do not guarantee successful system application or mission benefit.

Before proceeding with detailed discussions of its technology assessments, the committee summarizes these assessments in Table 2-1. The committee divides the major HS/H technology areas identified by it into constituent technologies and provides its assessment of (1) the current maturity of the constituent technology, (2) the impact of the constituent technology on the major technology area, (3) the likelihood that the technology will achieve the desired results, and (4) the overall criticality of each of the five major technology areas. Both the impact and the criticality evaluations are based on the committee's estimate of the anticipated contribution to the overall NAI goals—economics, safety, reliability, responsiveness, performance, and benefit to the industrial base. The impact was assigned descriptors according to the following (increasing) scale: negligible impact, little impact, moderate impact, significant impact, and extreme impact. Maturity levels are assigned numerical values corresponding roughly to the DoD S&T categories (not to be confused with TRLs, which are stated in parentheses): (1) basic research, (2) applied research, (3) advanced development, (4) demonstration/validation, and (5) engineering and manufacturing development.

TABLE 2-1 Status of Technologies for the High Speed/Hypersonics Pillar

NAI-Defined Technology Areas		Committee Evaluations			
		Constituent Technology Ratings			
Major Technology Area	Constituent Technology	Current Maturity Level (1 to 5) [TRL][a]	Relative Impact on Major Tech Area (Negligible, Little, Moderate, Significant, Extreme)	Likelihood of Achieving in Time Frame (Current Funding Level) (Low, Medium, High)	Relative Criticality of Major Technology Area (Low, Medium, High)
Air-breathing propulsion	Scramjet combustors	2.5 [4-6]	Extreme	High (small scale) Medium (large scale)	High
	Fuel injectors/flame holders	2.5 [4-6]	Extreme	High (small scale) Medium (large scale)	
	HC integrated flowpath	3 [5]	Extreme	High (small scale) Low (large scale)	
	H_2 integrated flowpath	2.5 [4-5]	Extreme	High (small scale) Low (large scale)	
	Engine control system	2.5 [4-6]	Significant	High	
	Propulsion-airframe integration	2.5 [3-6]	Extreme	Medium	
Engine materials	Metal combustor panels	3 [3-7]	Significant	High	High
	Cooled CMC panels	1.5 [2-4]	Extreme	Low	
	Cowl lip	2.5 [3-6]	Extreme	Medium	
	Injectors	2.5 [4-6]	Extreme	High	

(continued on next page)

TABLE 2-1 Continued

NAI-Defined Technology Areas		Committee Evaluations			
		Constituent Technology Ratings			
Major Technology Area	Constituent Technology	Current Maturity Level (1 to 5) [TRL][a]	Relative Impact on Major Tech Area (Negligible, Little, Moderate, Significant, Extreme)	Likelihood of Achieving in Time Frame (Current Funding Level) (Low, Medium, High)	Relative Criticality of Major Technology Area (Low, Medium, High)
	Seals	2 [3-4]	Extreme	Medium	
	Sensors	2.5 [3-5]	Significant	Medium	
Airframe materials	Al cryogenic tanks	2.5 [3-5]	Extreme	Medium	High
	Graphite-epoxy cryogenic tanks	1.5 [3-4]	Extreme	Low	
	Leading edges	3 [4-6]	Extreme	Medium	
	Thermal protection systems	2.5 [3-6]	Extreme	Medium	
	Structure	3 [3-6]	Significant	High	
Integrated vehicle design and MDO tools	Parametric geometry	3 [4-5]	Extreme	Medium	High
	Automated data transfer	3 [4-5]	Extreme	High	
	Automated grid generation	2.5 [2-5]	Extreme	Medium	
	Robust optimization	2 [3-4]	Extreme	Medium	

	Grid computing	3.5 [5]	Moderate	High	
	Uncertainty management	1.5 [2-3]	Extreme	Medium	
	High-fidelity HS/H surrogates	1.5 [1-5]	Extreme	Low	
Integrated ground test and numerical analysis	High-enthalpy ground test facilities	3 [5-6]	Extreme	Low	High
	Optical diagnostics	2.5 [5-6]	Significant	Low	
	Large eddy simulation methods	2 [4]	Extreme	Medium	
	Thermochemical models	2 [4-5]	Significant	Low	

[a] Current maturity levels, with a scale of 1 through 5, and technology readiness levels, with a scale of 1 through 9, are both used in various sections of this report. Both ratings are used in this table to enable the reader to evaluate technologies on both scales.

Technologies with low maturity levels, high impact, and low likelihood of achievement should be afforded special attention. Conversely, technologies with high maturity levels, low impact, and high likelihood of achievement should receive the same or less attention. Table 2-1 should help to determine if appropriate attention is being paid to the technologies supporting the NAI goals.

Air-Breathing Propulsion and Flight Demonstration

Propulsion is foremost among the critical technologies that will enable air-breathing hypersonic flight, not only for the dual-mode ramjet/scramjet engines that will achieve hypersonic speeds but also, although to a lesser degree, for the low-speed engines that will accelerate to ramjet/scramjet takeover speeds (typically Mach 3 to 4). Several challenges exist for scramjet engines, among them performance and operability across a broad range of operating speeds, especially above Mach 8; management of the extreme thermal environment encountered by the engines; and the durability of engine structures and systems for long life and low maintenance. Challenges for low-speed engines include the thrust-to-weight ratio (high thrust at low weight while maintaining high efficiency), thermal management, and integration with both the airframe and the high-speed propulsion system.

Technology development for scramjet engines has advanced at a slow but steady pace across several fronts since the 1960s; however, much work remains to be done (Curran, 2001). Substantial progress was made in scramjet design, materials, and performance during the National Aerospace Plane (NASP) program of the late 1980s and early 1990s (Waldman and Harsha, 1990), and progress continues. Between the NASP, NASA X-43A (McClinton et al., 2001), and Air Force HyTech (Powell et al., 2001) programs, the TRLs of scramjets and their associated components have been climbing since the mid-1980s. Current TRLs of hydrogen and hydrocarbon scramjet engines and their major components are listed in Table 2-2 (Bowcutt, 2003).

Many options exist for low-speed accelerator engines, but lately the bulk of the activity has focused on high-speed turbine engines and rocket-based combined cycle (RBCC) engines (Hueter and Turner, 1999), the latter being essentially a scramjet with rocket motors in the flow path for generating static and low-speed thrust. An added benefit of RBCC engines is that their embedded rocket motors can also generate exoatmospheric thrust for orbit insertion after the scramjets are shut down. NASA has pursued development of an RBCC engine for the past several years; although development was recently curtailed, it remains a viable propulsion system for access to space.

High-speed turbine engines, on the other hand, are being developed under the Air Force's Versatile Affordable Advanced Turbine Engines (VAATE) program (Lewis, 2003) and NASA's Revolutionary Turbine Accelerator (RTA) program (Bartolotta and McNellis, 2002; Bradley et al., 2002). Under these programs, steady progress is being made on small-scale and mid-scale engines that will operate at speeds up to Mach 3-4+. Current TRLs for high-speed turbine engines and their major components are listed in Table 2-2.

Development of both scramjets and high-speed turbine engines for space access applications may be feasible in the time frame of NAI—that is, by 2018. However, to hope to succeed in this goal, the nation must develop and demonstrate full-scale scramjet and low-speed propulsion systems and demonstrate their durability, efficiency, safety, reliability, and responsiveness, as well as their lower operating costs than current launch systems. Adequate funding must also be made available, timed to track the development spending profile.

Besides funding, other challenges must be tackled in order to accomplish propulsion development goals:

- Wind tunnel limitations that make it difficult to develop engine technologies and integrated systems and to validate engine performance at speeds above Mach 8 (see subsection on integrated ground test and numerical simulation and analysis).

TABLE 2-2 Current Technology Readiness Levels of High-Speed Turbine and Scramjet Engine Components and Systems

	Mach 0-4 Turbine Hydrocarbon	Mach 3-7 Hydrocarbon	Mach 3-7 Hydrogen	Mach 7-14 Hydrogen
Engine performance and operability[a]				
Inlet	5	5-6	5-6	5
Isolator	N/A	5-6	5-6	N/A
Fuel injectors/flameholders	4 (AB)	5-6	6	4-5
Combustor	6	5	5-6	4
Nozzle	4-5 (TMS)	5-6	4	4
Integrated flowpath	4 (including AB and TMS)	5	4	4
Structures and materials				
Cooled materials	5 (turbomachinery)	7-8	4	4
Uncooled materials	5	5	4	4
Cooling panels	4 (nozzle and combustor)	5-6	3	3
Variable geometry (e.g., seals)	4 to 5	4	3	3
Engine subsystems				
Sensors	4 to 5	6	6	3
Valves	N/A	5	5	4
Pumps	N/A	6	6	6
Active control system	4	4-5	6	5
Fuel to air heat exchanger	3 to 4	N/A	N/A	N/A

NOTE: Items in parentheses reflect requirements for a turbine-based combination cycle (TBCC) system. AB, afterburner; TMS, thermal management system.
[a] K.G. Bowcutt received inputs from the following: Chuck McClinton, NASA Langley; Robert Mercier, AFRL; Paul Bartolotta, NASA Glenn; Fred Billig, Pyrodyne (JHU/APL) (retired); Bill Imfeld, ASC (retired); Allen Goldman and George O'Connor, Boeing Rocketdyne; Steve Beckel, Pratt & Whitney.
SOURCE: Bowcutt, 2003.

• Numerical simulation—for example, computational fluid dynamics (CFD)—of engine flows and the integration of ground test and numerical simulation to counter wind tunnel limitations (see subsection on integrated ground test and numerical simulation and analysis).
• Integrated vehicle design methods that enable operable, high-performance integration of engines with airframes (see subsection on integrated vehicle design).
• Lightweight, high-temperature engine materials and thermal management technologies (see subsection on materials).
• Transition from a low-speed propulsion mode (e.g., a turbine) to a high-speed propulsion mode (e.g., a scramjet), especially when multiple flow paths and variable geometry are employed.

Availability of propulsion systems ready for full-scale development (FSD) by 2018 also requires thorough definition and planning of the fundamental research, component technology development, and ground testing and flight demonstrations required to mature full-scale engines to the point where they can be used for operational hypersonic flight.

Recommendation 2-10. The committee recommends that over the next 5 to 7 years, DoD and NASA, through NAI, continue to support programs aimed at developing and flight-testing small-scale hydrogen and hydrocarbon engines and begin additional activities to develop a mid-scale scramjet engine and demonstrate scramjet operation at the highest anticipated operational speeds (approximately Mach 14). For the long term, along the path to meeting the air-breathing hypersonic launch vehicle full-scale development objective, the committee recommends that engine development proceed by increasing engine scale in at least three steps: small-scale, mid-scale, and large-scale. Owing to ground testing limitations, each size engine will require testing in flight, so that flight demonstration vehicles will be required for each engine as well. Flight is also required to demonstrate engine mode transitions; to fully characterize engine and airframe thermal environments across the speed regime; to demonstrate the durability, effectiveness, and operational utility of integrated airframe-TPS and engine thermal management systems; and to validate the performance of highly integrated hypersonic vehicles designed using integrated analysis and multidisciplinary optimization techniques (see subsection on integrated vehicle design).

Recommendation 2-11. It is also recommended that the integration of critical components be continually increased (e.g., engine, then engine plus operational TPS, then engine plus integrated tank/airframe/TPS) on flight demonstration vehicles as engine scale increases. If requirements for, and definition of, the final large-scale flight demonstration are thoughtfully developed, then this demo vehicle might have some residual operational capability—as, for example, a CAV delivery system—which would be desirable in terms of getting the most value from the technology maturation investment.

Recommendation 2-12. In laying out this engine development and flight demonstration plan, it is recommended that many different engine and airframe concept options be explored so that no promising approaches are left undiscovered, thereby maximizing the chances of finding cost-effective air-breathing hypersonic solutions to civil and military access to space and various other military missions. A notional plan for propulsion development and flight demonstration along the lines described is shown in Figure 2-4.

From information available to the committee, it appears that sufficient funding is planned and available for near-term (i.e., small-scale) scramjet propulsion development and flight demonstration activities, from both the Air Force (at \$50 million to \$100 million per year) and NASA. An exception to this might be funding for the SED, due to congressional cuts made to NAI hypersonics elements of the 2004 Air Force S&T budget. Though unclear at this point, it is anticipated that funding levels in 2005 and beyond will make up for this 2004 shortfall. If this does not happen then adequate funding for SED should be secured beyond 2004. Another exception to adequate near-term funding is that associated with fundamental research, which is deemed to be insufficient at this time, with no known plans to correct the situation.

High-speed turbine engines appear to be funded by NASA at levels sufficient to develop small-scale and mid-scale ground demonstration engines in this decade under the RTA program. However, there is no known NASA plan to fund development of a large-scale engine that might be required for air-breathing access to space. The Air Force has a funding plan for developing and ground demonstrating a large high-speed turbine engine under the VAATE program, but funding is not available until 2006, and the amount available is less than planned and required.

For the long term, little information was provided to indicate the level of funding planned and/or available to execute the remainder of the NAI hypersonics roadmap. The amount needed to fund the complete roadmap to 2018 for large-scale scramjet and turbine engines, such as the notional one shown in Figure 2-4, is expected to be between \$3 billion and \$4 billion for engine ground development and another \$5 billion to \$10 billion for flight demonstration. This would require at least 10 times the current annual investment. Note that these estimates do not include the costs of developing other technologies critical to air-breathing hypersonic flight, such as tank, airframe, and TPS materials and structures. An indication that funding will not be sufficient for the long-term NAI plan can be seen in the NASA funding profile for the Next Generation Launch Technology program (i.e., NASA's contribution to the HS/H pillar of NAI) shown in Figure 2-5

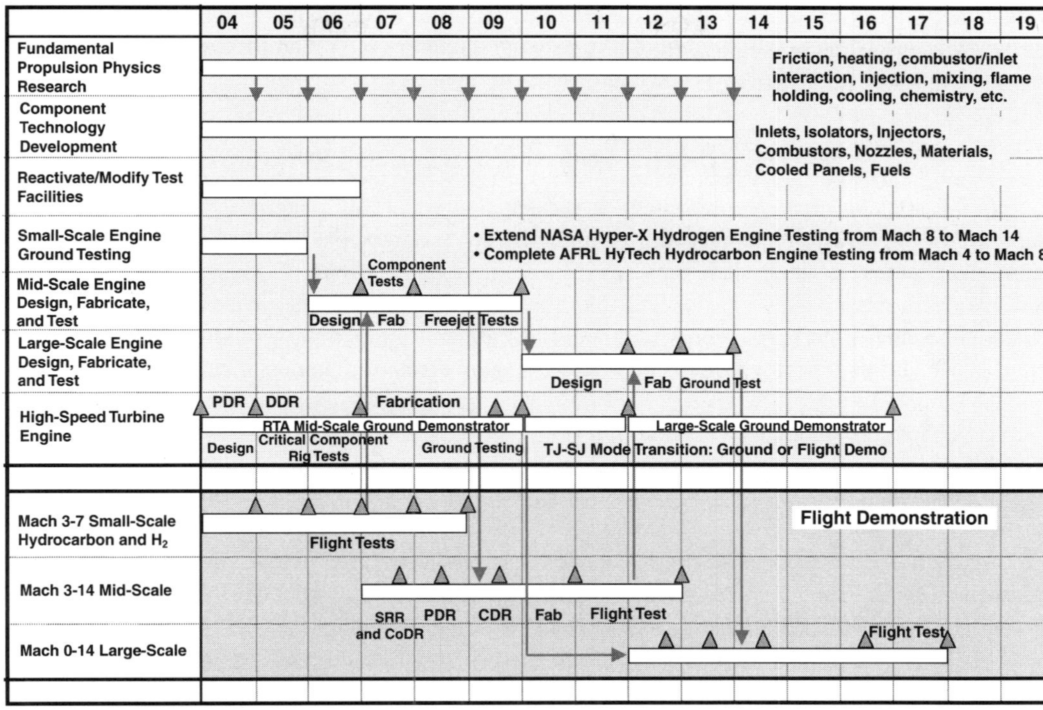

FIGURE 2-4 Notional hypersonic propulsion system development and flight demonstration roadmap. SOURCE: Adapted from Bowcutt, 2003.

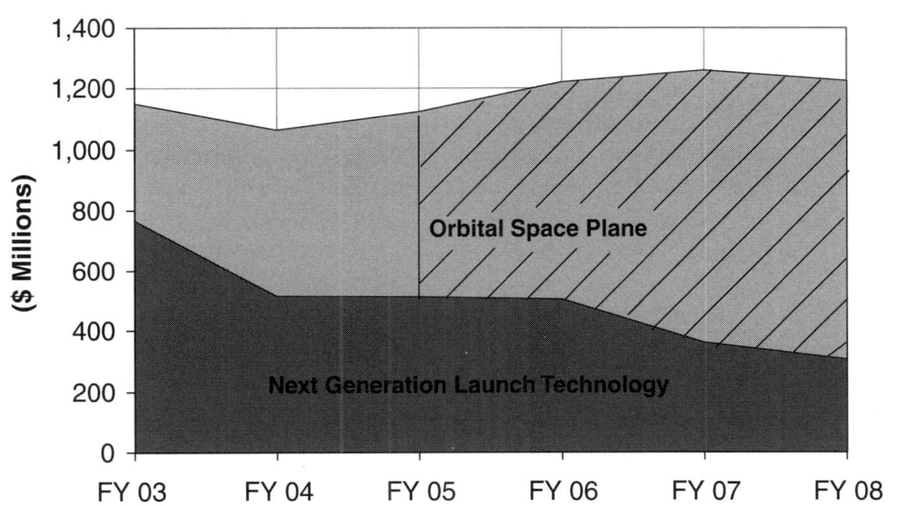

	FY03 BAU	FY 03	FY 04	FY 05	FY 06	FY 07	FY 08
Orbital Space Plane	295.7	383.3	550.1	609.9	716.6	894.7	916.0
Next Generation Launch Technology	583.7	767.1	514.5	513.8	504.7	362.3	307.4
Space Launch Initiative Total	879.4	1,150.4	1,064.6	1,123.7	1,221.3	1,257.0	1,223.4

FIGURE 2-5 Space Launch Initiative full cost budget (FY 2003 reflects estimated full cost). SOURCE: Rogacki, 2003.

(Rogacki, 2003; Lyles, 2003), where declining budgets after FY 2006 are evident. Clearly, significantly increased budgets will be required to execute the entire NAI plan to 2018, and the funding will have to come from the Air Force and/or NASA.

Materials, Thermal Protection Systems, and Structures

NASA and Air Force representatives presented the state of materials development to the committee. They made clear that although significant progress has been made in recent years, additional progress must be made rapidly if the schedule laid out in the NAI program is to be kept. Because of the broad array of structural requirements, heat loads, and environments in a typical hypersonic vehicle and propulsion system, a full treatment of this topic would require a dedicated report. Instead, this report will focus on three important areas of materials technology, as follows:

- Thermal protection systems (TPSs),
- Actively cooled combustor panels, and
- Cryogenic tanks.

The viability of hypersonic flight vehicles depends, to a large extent, on the availability of lightweight TPSs. Such systems may be either cooled (active) or uncooled (passive), and both ceramic and metallic TPS approaches have been successfully applied.

Most of the windward surface of the space shuttle orbiter is subjected to moderately high heat loads during reentry, and it is protected by silica foam insulation tiles (Korb et al., 1981). Although these tiles have proven to be reliable in service from a safety perspective, they are extremely fragile and subject to damage from impact by rain, ice, and other objects. They also require frequent recoating and repair and are therefore not suitable for use where rapid response is a requirement.

Several approaches have been taken toward designing a more rugged passive TPS system. One of the more promising uses a Nextel-fiber-toughened outer layer bonded to conventional silica foam tiles using a monazite ($LaPO_4$) powder binder (Davis et al., 2000). The resulting alumina-monazite ceramic composite has been demonstrated to provide the necessary toughness and damage tolerance for an impact-resistant thermal protection system.

Metallic TPSs have also been investigated and tested, but the operational experience is much less extensive than with the ceramic and ceramic composite systems. One advantage of the metallic TPS approach is that active cooling can be incorporated for the highest heat flux areas, such as leading edges and nose cones.

In the propulsion system of a hypersonic vehicle, active cooling of certain components such as the cowl lip and combustor panels will be necessary. Although extensive testing of various candidate designs has been accomplished, no single solution that solves all the operational requirements has emerged (Sillence, 2002). In the combustor section of a hypersonic propulsion system, active cooling is necessary, even at modest Mach numbers. Figure 2-6 shows the wall temperature of uncooled combustors (the adiabatic wall temperature) as a function of Mach number, along with the maximum use temperature of a variety of candidate combustor materials (Faulkner et al., 2003).

Metallic combustor panels using modifications of existing rocket combustion chamber designs have the appeal of considerable operational experience but tend to be too heavy. Recently, several ceramic matrix composite (CMC) designs have been fabricated and tested (McClinton, 2003). Although some of these concepts show significant promise, none completely satisfy the requirements and all need significant additional effort to reduce risk to the point where they could be incorporated into an operational hypersonic propulsion system.

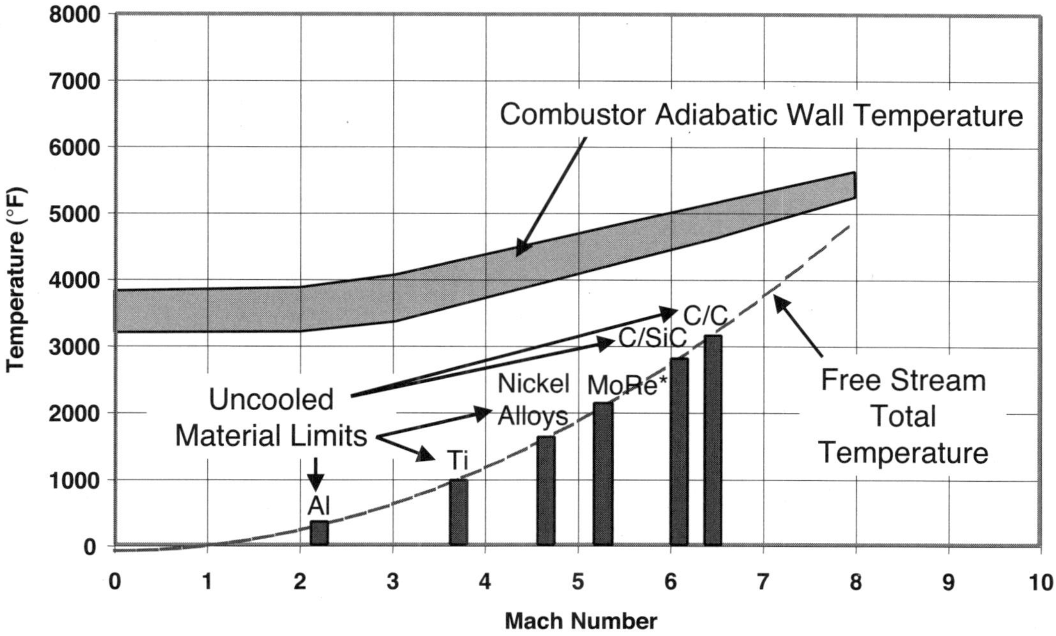

FIGURE 2-6 Combustor wall temperature as a function of Mach number. Active cooling is required in the combustor. Asterisk denotes theoretical structural limit; oxidation limit is much lower. SOURCE: Faulkner et al., 2003.

Cryogenic tanks have received considerable attention in recent years, and both metallic and composite designs that meet design criteria are available. The development of friction stir welding (FSW) technology has greatly improved the producibility and quality of advanced aluminum tanks, while advances in the understanding of large composite structures' manufacturing has improved the efficiency and integrity of composite tanks. The remaining step is to produce and flight test a fully reusable, full-scale test article through multiple cycles (Chase et al., 2002).

For materials, TPSs, and structures, air-breathing approaches to hypersonic speeds are far more demanding than rocket-based approaches owing to sustained, high-aerodynamic-heating environments, particularly above Mach 8. There seems to be a validation gap between numerous new high-temperature materials and their confident application. Further, NASA seems to be much more optimistic than the Air Force about the technological readiness of many of these materials. This suggests the NAI program needs

- Better coordination and agreement between the DoD and NASA on the definition of material readiness and
- A reinvigorated basic and applied research budget to move these materials further along in their TRLs.

The feasibility of achieving adequate TRLs in the near term appears good for a Mach 8 missile. As the vehicle size increases, the risk increases appreciably, particularly at the higher Mach numbers. Table 2-3 quantifies this in terms of committee TRL estimates.

Explicit estimates of the funding needed depend to a large extent on vehicle Mach number. To attain Mach 8 in the near term on a scale suitable for a missile application with an acceptable level

TABLE 2-3 Estimates of Current TRL Status for a Number of Critical Airframe and Engine Materials

	Mach 3-7 Hydrocarbon	Mach 3-7 Hydrogen	Mach 7-14 Hydrogen
Engine materials			
Combustor panels (metallic)	7 (6)	5 (4)	4 (3)
Cooled CMC panels	4 (3)	3 (2)	3 (2)
Cowl lip	5-6	4	3
Injectors	5-6	5-6	4
Seals	4	4	3
Sensors	5	4	3
Airframe materials			
Cryogenic tanks (Al)	5 (4)	5 (4)	4 (3)
Cryogenic tanks (Gr-Ep)	4 (3)	3	3
Leading edges	5-6	5-6	4
TPS	8 (7)	8 (7)	7 (6)
Structure	6 (5)	6 (5)	4 (3)

NOTE: Numbers in parentheses are for large-scale, reusable applications.
SOURCE: McClinton, 2003; Bowcutt, 2003.

of risk, a modest increase in the present investment levels is probably adequate. However, to attain Mach 15 in a large-scale vehicle by 2018 would probably require increasing the materials technology investment by a factor of 5 to 10, resulting in a total investment level of $150 million to $200 million per year.

Integrated Vehicle Design and Multidisciplinary Optimization

Integrated vehicle design and multidisciplinary optimization have been identified by the committee as a critical technology for the HS/H pillar of the NAI.

Air-breathing hypersonic vehicles will consist of highly integrated systems. At hypersonic Mach numbers, much of the airframe must act as the inlet and nozzle for the propulsion system. The airframe must also mitigate the effects of large propulsive lift and pitching moments and large Mach and dynamic pressure variations in flight. The shape of the vehicle will determine the vehicle structure, the type of integrated thermal protection system and its material, the control system, and the flight mechanics and trajectory. The flight trajectory will in turn determine the aerodynamic heating loads that will influence vehicle aeroelastic behavior, aeropropulsive performance, TPS, and hence empty weight. The empty weight of the vehicle is also directly related to the structural shape, and the vehicle airframe will affect the fuel and payload volume since, unlike conventional aircraft, the majority of the volume in a hypersonic vehicle must accommodate fuel. The application of multidisciplinary design optimization (MDO) is vital to obtaining robust vehicle designs that satisfy all constraints, including off-design performance. MDO can also allow the designer to investigate systematically the complex interaction of design trade-offs between the necessary disciplines, changes to the objective function, the sensitivity of the design to various parameters, and uncertainty.

MDO in engineering design is essentially the solution of single or multiple optimization problems constructed from a coupled system of single-discipline, black-box legacy computer codes as opposed to the solution of a single tightly coupled code. MDO objective functions and constraints depend not only on optimization variables but also on ancillary variables such as solutions to coupled systems of stand-alone solvers as surrogates—for example, discipline-specific partial differential equations, table look-ups, or other nonsmooth analysis codes. This has important implications for optimization since the user must be concerned with how data are transferred between surrogates and which data are transferred. In addition, the objective function and constraint values may be expensive to compute and may be nondifferentiable and discontinuous. Adding to the difficulties, many nonlinear programming techniques require that the user open the single-discipline solvers in order to evaluate internal parameter derivatives, if they exist. This can be costly and is generally resisted by the single-discipline specialist, though progress is being made in providing this necessary information via automatic differentiation.

When applied to typical engineering problems, MDO will give a point solution—that is, the solution that satisfies an objective function and set of constraints. However, this solution is rarely the one that aircraft designers need. There is almost always more than one objective involving a mixture of continuous, discrete, and categorical parameters, and the objectives often conflict with one another. The solution is usually one that comes from studying the trade-offs between these competing objectives. It must also satisfy the problem constraints and give a robust estimate of the likely value of the objective function in the actual design. Moreover, the aircraft designer will often choose computational efficiency over finding a better solution, but obtaining efficiency is especially problematic when dealing with fluid dynamics. For example, varying the parameter values requires regridding the spatial domain around configurations and rerunning computational fluid dynamics (CFD) solvers. The effort expended to do this is dependent on whether the CFD solver uses a structured or an unstructured grid and on the degree of fidelity of the fluid flow equations being solved: Euler vs. Navier-Stokes, for example. Efforts are being made to automate the layout for both structured and unstructured grids. The model fidelity issue is sometimes addressed by using low-fidelity surrogates at the start of the optimization and increasing the fidelity as the optimal solution is approached, a procedure called management of variable fidelity surrogates. A method used to address the issue of fidelity is simply to increase the computational resources through inexpensive supercomputing or grid computing. In any of these approaches, determining the effect of a design parameter adjustment as it propagates through all of the surrogates is not straightforward. Arguably, the most critical hurdle in conventional aircraft design is the understanding, tracking, and management of uncertainty in and between surrogates within the MDO framework.

In addition to the previously mentioned weaknesses in applying MDO to conventional aircraft design, the hypersonic vehicle designer must be able to construct high-fidelity fluid dynamics surrogates in the hypersonic regime. The construction of high-fidelity hypersonic fluid dynamic surrogates will require the following:[3]

- Improved understanding and modeling of shock and turbulent boundary layer interactions,
- Measurements of nitrogen oxide and other species to support high-enthalpy wind tunnel studies,
- Accurate modeling of finite-rate chemistry effects on turbulent boundary layers,
- Validation and verification (V&V) of boundary layer transition models, including the effects of finite-rate chemical reactions,
- V&V of Reynolds-averaged Navier-Stokes (RANS) models for hypersonic flows, including flows with large favorable and adverse pressure gradients,
- V&V of finite-rate chemistry models, since the limits of kinetics models used for air, hydrocarbon-air, and hydrogen-air are not well known, especially under thermal nonequilibrium conditions,

[3]Graham Candler, University of Minnesota, personal communication to committee member Kevin Bowcutt in October 2003.

TABLE 2-4 Committee Estimates of the Current Technology Readiness Levels of MDO Applied to High Speed/Hypersonics Vehicle Design

MDO Components	TRL	Years to a TRL of 6
Nonlinear optimization	6-7	0
Automatic differentiation	6-7	0
Mixed-variable optimization	5	2
Grid computing	5	2
Automated data transfer between surrogates	4-5	5-7
Parametric geometry	4-5	5-7
Robust solutions through optimization	3-4	5-7
Automated grid generation	2-5	5-7
Management of variable-fidelity surrogates	2-3	5-7
Uncertainty management	2-3	5-7
High-fidelity HS/H surrogates	1-5	10-15
V&V of HS/H surrogates	1	10-15

- Development of hybrid direct simulation Monte Carlo–Navier-Stokes (DSMC–NS) solvers for high-altitude aerodynamics, and
- Development of large-eddy simulation and discrete-eddy simulation (LES/DES) methods for highly compressible flows, including shock waves, fluid mixing, and chemical reactions.

Current TRLs of the MDO components applied to aircraft design vary from high to low. Processes that are well understood and mature include nonlinear optimization techniques and automatic differentiation. Of moderate readiness are mixed variable optimization and grid computing. Automated data transfer between discipline-specific surrogates can be done regularly through the commercially available ModelCenter code[4] and the publicly available Design Analysis Kit for Optimization and Terascale Applications (DAKOTA) code.[5] Immature and low-TRL components include the determination of robust solutions through optimization, the efficient propagation of design parameter changes with faster automated regridding, management of variable-fidelity surrogates, and uncertainty management of the discipline-specific surrogates. The components of lowest TRL are high-fidelity HS/H surrogates and the V&V of these surrogates. The TRL for each of the MDO components is given in Table 2-4. The displayed TRL estimates are the product of this NRC committee only and are not based on government sources.

Numerous researchers around the country are successfully addressing these challenges in MDO. Researchers include staff of NASA Langley,[6] the AFRL,[7] Sandia National Laboratories' Advanced Simulation and Computing program,[8] Los Alamos National Laboratory, and various universities around the country. The committee believes the number of years to a TRL of 6 shown in Table 2-4 will hold if these groups collaborate and if NASA's uncertainty management and MDO efforts are funded over the next 10 years to a total of at least $200 million. It must be emphasized that the problems specific to hypersonic vehicle design are long term and will require several more years of adequate and sustained funding at a level of several tens of millions more dollars to fully solve.

[4]Accessed at http://www.phoenix-int.com/products/ModelCenter.html.
[5]Accessed at http://endo.sandia.gov/DAKOTA/software.html.
[6]Accessed at http://mdob.larc.nasa.gov.
[7]Accessed at http://www.va.afrl.af.mil/COE/MDT/mdt_index.html.
[8]Accessed at http://www.nnsa.doe.gov/asc/program_overview.htm.

Integrated Ground Test and Numerical Simulation/Analysis

Introduction

Success in both external hypersonic flow and in air-breathing propulsion is critically dependent on the integration of ground testing (GT) and CFD for a number of reasons. At Mach numbers greater than, say, 8, the stagnation temperature is greater than 3000 K, and above Mach 12, greater than 7000 K. Materials from which ground test facilities are made cannot sustain associated heat loads for extended periods, so that short-duration methods such as shock tunnels and expansion tubes must be resorted to. Effects of prolonged heat loads on materials and engine control experiments are therefore inaccessible to such facilities. Also, only some of the relevant dimensionless parameters can be duplicated exactly in any one facility. The most successful experiments involving air-breathing combustion have, to date, been so-called direct-connect experiments, and accurate whole-model drag measurements in ground test facilities at high enthalpy are nonexistent. An urgent need in all high-enthalpy, short-duration facilities is adequate characterization of the free-stream conditions. In particular, the chemical composition and translational and vibrational temperatures are inadequately known. Diagnostic methods for measuring instantaneous planar distributions of particular chemical species concentrations and vibrational and translational temperatures do exist, but they are expensive and difficult to apply to large facilities. The need for such measurements applies equally to test article flow fields, of course, and is therefore not restricted to free-stream conditions.

Computational techniques, while largely successful in many low-Mach-number flow applications, where high-enthalpy or finite-rate effects, and even high-compressibility effects, do not cause problems, suffer severely when the Mach number exceeds 8 or so. Chemical reaction rates have often only been measured at temperatures lower than those encountered in practice, so unjustified extrapolation is generally used. Furthermore, rates depend on the state of nonequilibrium that exists, and very little is known about that dependence. Coupling between vibrational excitation and dissociation can cause rates to be significantly changed, and models for these processes have not been adequately validated. Similarly, models for surface chemistry and the chemistry of hydrocarbon combustion are in need of validation. Large-eddy simulation of highly compressible, reacting turbulent flows involving shock waves is almost entirely undeveloped.

The number of parameters for a vehicle development program is increased very substantially by increasing the flight Mach number from, say, 3 to 8 or more. This makes it impractical to employ flight testing for exploratory purposes.

The philosophy for making significant progress must therefore be as follows:

- Characterize the free-stream conditions of short-duration facilities as accurately as possible, using modern diagnostic tools.
- Perform benchmark experiments in well-characterized flows, again using sophisticated modern diagnostics. Perform identical experiments in different facilities.
- Compute the benchmark experiments with existing numerical techniques to validate or establish doubts about the techniques.
- Use validated techniques for conditions that lie off or between benchmark conditions for development work (including different scales and different flow conditions).

At the same time, it is most important that improved computational methods be developed, taking advantage of Moore's law rates of increase in computational power, and that the most effective diagnostic methods be selected and new ones developed in appropriately funded 6.1 programs. The most effective funding for research of this type is presently provided by the Air

Force Office of Scientific Research (AFOSR); however, AFOSR's annual budget has decreased (in inflation-adjusted dollars), from $456 million in 1965 to $218 million in 2000, while the number of topics supported has increased significantly. NASA resources are similarly small and, though apparently focused, are in fact quite heterogeneous because of the large number of principal investigators in each effort.

Many of the scientific aspects of the challenges in computational methods are the same ones that also face scientists in the national laboratories of the DOE. It is therefore worth exploring collaborations with these persons and labs to leverage funding sources.

Existing and Projected Ground Testing Facilities

The facilities existing in the United States for testing at enthalpies corresponding to Mach numbers greater than 10 include the shock tunnels and the expansion tunnel at the Calspan-University of Buffalo Research Center, Inc. (CUBRC); the T5 hypervelocity shock tunnel at the California Institute of Technology; the expansion tube at the General Applied Science Laboratory (GASL); and the G-Range reflected shock tunnel at the Arnold Engineering Development Center (AEDC). All of these are short-duration facilities operating in the test-time range from 1 to 10 milliseconds, depending on enthalpy, with the shorter times corresponding to higher enthalpy. All of these facilities (except the G-Range) have been extensively used for both fundamental research and industrial testing.

Characterization of the flows in these facilities is not sufficiently complete. The most detailed characterization has been done in one of the shock tunnels at CUBRC by comparing computational predictions of the tunnel nozzle flow and the flow over a sensitive shock-wave–boundary-layer interaction model with experimental measurements (Candler et al., 2002). Information about the composition and vibrational excitation of the free stream in all the facilities is indirect, via comparisons of experimental and computed flows over models using surface heat flux, interferometry, or pressure measurements. Existing optical diagnostic techniques should be used to measure composition and vibrational excitation directly (Bessler et al., 2002; Ben-Yakar and Hanson, 2002; Rossmann et al., 2002). The lack of such data is directly caused by the high cost of running the facilities and the high cost (or development cost) and sophistication of the diagnostics. Detailed information about flow noise is also lacking.

The larger size of the facilities at CUBRC makes them more suitable for industrial testing. Also, the expansion tunnel at CUBRC is a facility in which effective reservoir pressures up to 600 MPa can be reached. The T5 at Caltech has mainly been used for high-enthalpy fundamental research, which has discovered important, previously unknown effects.

An ongoing effort by Princeton University is to work toward a facility in which high-pressure air is first expanded to a low supersonic Mach number in a nozzle, then heated by magnetically controlled electron beams, and subsequently expanded to a high Mach number in the same nozzle. This device, if successful, would produce a hypersonic flow with virtually pure air and very low nitric oxide concentration for a projected run time of 10 seconds. So far, elements of this process have been demonstrated at below projected power, but the facility is very far from completion, and many details of the principle of operation are still uncertain. The facility may not be operational for at least 10 years.

Facilities suitable for testing material properties under high heat load need to run for extended periods to give the heat time to diffuse into the solid. For this purpose arc tunnels provide a compromise. In such tunnels the gas is heated by an electric arc in the reservoir region of a continuous expansion through a water-cooled nozzle. Though the flow is continuous, the pressure in such facilities is much lower than in shock tunnels, so the free stream is much more highly dissociated and the flow field not as faithfully simulated. However, surface heat flux can be approximately duplicated for sufficiently long times to test materials. The largest such facility in the United States is operated at AEDC.

Longer test times (seconds rather than milliseconds) are also required for engine control development and investigation of aeroelastic phenomena. No ground test facilities exist at this point that can provide the necessary capability at Mach numbers greater than 8.

Uncertainties in Computational Methods

Some of the uncertainties in the application of computational methods to high-Mach-number flows have nothing to do with the algorithms employed. Rather they stem from the fact that the chemical reaction rates needed both for the dissociation and recombination in air and for combustion are not sufficiently precisely known. In particular, the reaction rates for gases in vibrational nonequilibrium (vibration-dissociation coupling) are subject to considerable uncertainty.

The same applies to surface chemistry. Above Mach 14, oxygen dissociates, and surface recombination must be considered for all TPS material and integrated into computations. Neglect of this effect could result in insufficient thermal margin in the design of hypersonic vehicle thermal protection systems (Jumper, 1995; Jumper and Seward, 1994).

The prediction of transition from laminar to turbulent boundary layer flow is critical to the design of hypersonic vehicles. While computational techniques have not succeeded in making predictions of transition at high Mach numbers, linear stability calculations combined with experiment have yielded insights showing that vibrational excitation can strongly influence the path toward, and therefore the location of, transition (Johnson et al., 1998; Adam and Hornung, 1997; Stuckert and Reed, 1994). However, much more work is needed in this area to gain sufficient confidence for design.

Computational techniques need particular efforts in the simulation of turbulence. In particular, LES techniques for high-enthalpy reacting flows must be developed in order to improve predictions of important quantities in turbulence.

Roadmap

In accordance with the philosophy outlined earlier, the roadmap for combined research in ground testing and computation should proceed along the following lines, which can be related to the notional timeline shown in Figure 2-4.

1. One purpose of combining ground testing with computation is to produce validated computational tools that can subsequently be used in design. To get there, identical benchmark experiments must be performed in the different facilities. The free stream of the facilities must be characterized accurately by combining modern diagnostic measurements with computation. Computations of the benchmark flows have to be made and compared with the data from experiments. Time required, 7 years; cost (facility upgrades and characterizations, benchmark experiments, computation), $80 million.

2. For each of the vehicles in NAI, external flow and integrated airframe-propulsion testing at the correct enthalpy and covering the parameter space is necessary in combination with application of validated tools to gain more confidence. The facilities most suited for this work are the AEDC tunnels in the Mach 5 to 8 regime and the CUBRC shock tunnels above Mach 8. Time frame, 2006-2015; cost, ~$1 million per vehicle per test.

3. Full-scale engine testing is not possible in currently existing ground test facilities, especially above Mach 8, but missile-scale engines can and should be directly ground tested at all designed-for operational speeds. Engines for hypersonic vehicles larger than missiles, on the other hand, can only be tested at subscale. Properly validated analysis tools must therefore be used to

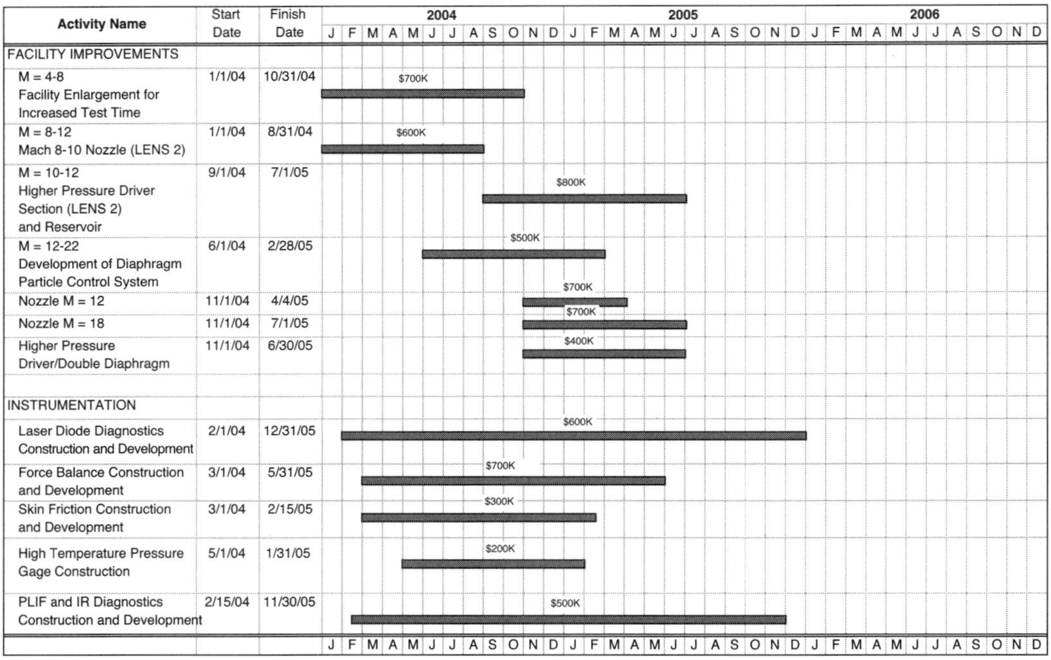

FIGURE 2-7 Perceived facility and instrumentation upgrade needs and costs ($7 million over 2 years). SOURCE: Holden, 2003.

supplement engine ground testing. The time and cost profiles are contained in the subsection on propulsion.

4. A 10-year research program on hypersonic flows, emphasizing high-enthalpy effects on each of the following:

—Transition from laminar to turbulent boundary layer flow,
—Turbulent boundary layers and shear flows,
—Shock wave-boundary layer interaction,
—Fuel injection, mixing, and reaction with air, and
—Integrated airframe-propulsion performance and operability.

Full advantage should be taken of modern optical diagnostics and computation for all of these. Time, 10 years; cost, $50 million.

Technology Readiness Levels

TRLs in the integrated ground test and numerical simulation/analysis area are as follows:

- High-enthalpy ground test facilities (subject to earlier described limitations): 5-6 (see Figure 2-7 for perceived upgrade needs and costs),
- Optical diagnostic methods: 5-6,
- Computational methods: 4-6 (high-enthalpy LES, 4; others higher),
- Chemical reaction rates at nonequilibrium conditions: 4-5, and
- Surface chemistry: 4-5.

REFERENCES

Published

Adam, P.H., and H.G. Hornung. 1997. Enthalpy effects on hypervelocity boundary-layer transition: Ground tests and flight data. J. Spacecraft and Rockets 34(5): 614-619.

Bartolotta, P., and N. McNelis. 2002. RTA-Demonstrating TBCC Technologies. 1st AIAA/IAF Symposium on Future Reusable Launch Vehicles. Huntsville, Ala. April.

Ben-Yakar, A., and R.K. Hanson. 2002. Characterization of expansion tube flows for hypervelocity combustion studies. J. Propulsion and Power 18(4): 943-952.

Bessler, W.G., C. Schulz, T.Lee, J.B. Jeffries, and R.K. Hanson. 2002. Quantitative NO-LIF imaging in high-pressure flames. Appl. Phys. B-Lasers O 75(1): 97-102.

Bradley, M.K., K.G. Bowcutt, J.G. McComb, P.A. Bartolotta, and N.B. McNelis. 2002. Revolutionary Turbine Accelerator (RTA) Two-Stage-to-Orbit (TSTO) Vehicle Study. AIAA Paper 2002-3902. 38th Joint Propulsion Conference. July 7-10.

Candler, G.V., I. Nompelis, M.C. Druguet, M.S. Holden, T.P. Waldhams, I.D. Boys, and W.L. Wang. 2002. CFD validation for hypersonic flight: Hypersonic double-cone flow experiment. AIAA Paper 2002-0581.

Chase, R., and Ming Tang. 2002. The quest of single stage earth-to-orbit: TAV, NASP, DC-X and X-33 accomplishments, deficiencies and why they did not fly. AIAA Paper 2002-5143.

Curran, E.T. 2001. Scramjet engines: The first forty years. Journal of Propulsion and Power 17(6): 1137-1192.

Davis, J.B., D.B. Marshall, P.E.D. Morgan, K.S. Oka, A.O. Barney, and P.A. Hogenson. 2000. Fuzzy logic control of structural vibration of beams. AIAA Paper 2000-0172.

Hueter, U., and J. Turner. 1999. Rocket-based combined cycle activities in the Advanced Space Transportation Program Office, AIAA 99-2352, 35th AIAA/ASME/SAE/ASEE Joint Propulsion Conference, Los Angeles, June 20-24.

Johnson, H.B., T. Seipp, and G.V. Candler. 1998. Numerical study of hypersonic reacting boundary layer transition on cones. Phys. Fluids 10(10): 2676-2685.

Jumper, E.J. 1995. Recombination of oxygen and nitrogen on silica-based thermal protection surfaces: Mechanism and implications. NATO ASI Molecular Physics and Hypersonic Flows. Maratea, Italy, May-June 1995.

Jumper, E.J., and W.A. Seward. 1994. Model for oxygen recombination on reaction-cured glass. Journal of Thermophysics and Heat Transfer 8 (July-September): 460-465.

Korb, L.J., C.A. Morant, R.M. Calland, and C.S. Thatcher. 1981. The shuttle orbiter thermal protection system. Amer. Cer. Soc. Bulletin 60(11): 1188.

McClinton, C.R., R.R. Reubush, J. Sitz, and P. Reukauf. 2001. Hyper-X program status. AIAA Paper 2001-1910. AIAA/NAL/NASDA-ISAS 10th International Space Planes and Hypersonic Systems and Technologies Conference, Kyoto, Japan, April.

Powell, O.A., J.T. Edwards, R.B. Norris, and K.E. Numbers. 2001. Development of hydrocarbon-fueled scramjet engines: The Hypersonic Technology (HyTech) program. Journal of Propulsion and Power 17(6): 1170.

Rossmann T., M.G. Mungal, and R.K. Hanson. 2002. Evolution and growth of large-scale structures in high compressibility mixing layers. J. Turbulence Vol. 3, article no. 009.

SAIC (Science Applications International Corporation). 1995. National aero-space plane advanced technology impacts, Final Report.

Sillence, M.A. 2002. Hydrocarbon scramjet engine technology flowpath component development. AIAA Paper 2002-17-5158.

Stuckert, G., and H. Reed. 1994. Linear disturbances in hypersonic chemically reacting shock layers. AIAA J. 32(7): 1384-1393.

Waldman, B.J., and P.T. Harsha. 1990. The National Aero-Space Plane program. AIAA Paper 90-5252.

Unpublished

Bowcutt, K. 2003. Hypersonic Technology Status and Development Roadmap. Briefing by Kevin Bowcutt, Chairman, Boeing Technical Fellowship Advisory Board, to the Committee on the National Aerospace Initiative. September 3.

Evans, D. 2003. National Aerospace Initiative: Materials and Processes. Briefing by Daniel Evans, AFRL, to the Committee on the National Aerospace Initiative. September 4.

Faulkner, R.F., and R. Kazmar. 2003. Pratt and Whitney Hypersonics Overview. Briefing to the Committee on the National Aerospace Initiative on the occasion of its site visit. November 7.

Graff, G. 2003. HyFly. Briefing by Gil Graff, Office of Naval Research, to the Committee on the National Aerospace Initiative. October 7.

Hickman, R., G. Keether, D. Boudreaux, P. Stewart, and G. Hernandez. 2003. Operationally Responsive Spacelift. Briefing by Robert Hickman, The Aerospace Corporation, to the Committee on the National Aerospace Initiative. October 7.

Holden, M. 2003. Facility and Implementation Improvements. Prepared by Mike Holden, CUBRC, in response to questions by committee member Hans Hornung.

Lewis, T. 2003. The Role of Turbine Engines in NAI: AFSTB NAI Review. Briefing by Timothy Lewis, AFRL, to the Committee on the National Aerospace Initiative. October 7.

Lyles, G. 2003. NGLT Briefing to the Air Force Science and Technology Board. Briefing by Garry Lyles, NASA Marshall, to the Committee on the National Aerospace Initiative. October 7.

McClinton, C. 2003. NASA Hypersonics. Briefing by Charles McClinton, NASA Langley, to the Committee on the National Aerospace Initiative. September 4.

Morrish, A. 2003. DARPA Programs for the National Aerospace Initiative Objective. Briefing by Art Morrish, DARPA, to the Committee on the National Aerospace Initiative. August 5.

Rogacki, J. 2003. The National Aerospace Initiative—A Synergistic Partnership between NASA and DoD. Briefing by John "Row" Rogacki, NASA, to the Committee on the National Aerospace Initiative. August 5.

Sega, R. 2003a. National Aerospace Initiative—NRC Briefing. Briefing by Ronald M. Sega, DDR&E, to the Committee on the National Aerospace Initiative. August 6.

Sega, R. 2003b. National Aerospace Initiative—NRC Briefing. Briefing by Ronald M. Sega, DDR&E, to the Committee on the National Aerospace Initiative. November 10.

Walker, B. 2003. National Aerospace Initiative and the U.S. Army. Briefing by Billy Walker, U.S. Army Aviation and Missile Command, to the Committee on the National Aerospace Initiative. August 6.

3

Access to Space

CURRENT OPERATIONAL CAPABILITIES

Access to space is an international capability, with many of the world's countries possessing very capable vehicles. A partial list of countries and their vehicles follows:

United States	Atlas, Delta, Pegasus, the space shuttles of the Space Transportation System, Taurus, and Titan
France	Ariane
Russia	Proton and Soyuz
Ukraine	Zenit
China	Long March
Japan	H-2

Some of these rockets can lift about 18,500 kg to low Earth orbit (LEO), and several are crew-rated. A space shuttle can put about 29,500 kg into a due east LEO. Raw lift performance is not the issue. The issues are cost, reusability, reliability/availability, responsiveness, and turnaround time to put assets into space.

The U.S. space launch capability has evolved over nearly 50 years in response to the requirements of military, intelligence, civil space, and commercial users. The U.S. military and intelligence communities have required relatively low launch rates of uncrewed spacecraft to a variety of operational orbits, ranging from low Earth to geosynchronous. Civil space users have also required only low launch rates. So far, commercial users have implemented only uncrewed spacecraft at low launch rates into Earth orbits.

The proliferation of terrestrial fiber optics and increasingly capable and longer life satellites (both commercial and military) have tended to reduce launch rates, thereby inhibiting large new investments in launch vehicles and associated infrastructure. In recent years, fewer commercial launches and foreign (often subsidized) competition have made domestic uncrewed launch suppliers much less economically viable than had earlier been predicted (Antonio, 2003).

However, partly owing to the planned evolution of expendable launch systems by the Air Force, the U.S. government currently has available three families of expendable rockets: the soon-to-be-retired Titan IV and the about-to-be-flown Delta IV Heavy for heavy lift, and the old and new families of Delta and Atlas vehicles for medium lift. In addition, two small launchers (Taurus and Pegasus) provide on-orbit delivery of lighter spacecraft. These systems are completely expendable, will support only low launch rates, and are not crew-rated. The medium- and heavy-lift launchers also operate exclusively from two highly vulnerable Florida and California coastal locations, and one of the boosters (the Atlas V) incorporates a Russian-designed and -manufactured liquid main engine.

Our nation has relied for more than 20 years on the only crew-rated reusable launch vehicle (RLV) in the world, the Space Transportation System (the shuttles). Now, two catastrophic failures in 113 flights have reduced the shuttle fleet to three, which are grounded until corrective actions can be implemented, and have severely limited the shuttle manifest. Currently, crew access to the International Space Station (ISS) depends on the Russian space program to provide limited capability with expendable launch vehicles and crew modules. If the ISS is to be completed and to have the full crew complement, the space shuttle will have to be brought back into service or there will have to be an extended delay until a replacement vehicle is available.

This chapter contains at various points 34 findings and 26 recommendations.

PLANNED CAPABILITY

Both the Air Force and NASA are beginning to pursue new capabilities to satisfy their respective mission needs. While potential new systems may differ considerably and are not yet fully defined, many of the underlying technologies are expected to be similar. As stated earlier, raw lift performance is not the issue. The issues are cost, reusability, reliability/availability, responsiveness, and turnaround time to put assets into space. NAI has targeted these issues in its access-to-space (ATS) pillar with a three-phase program of increasingly demanding requirements. Vehicle and supporting system characteristics and, by implication, NAI technology timelines are illustrated in Figure 3-1.

The ATS pillar has identified the common technical efforts and provides a mechanism for cooperation, sharing, and advocacy of these technologies. Included are reusable rocket propulsion, tanks, and airframes; thermal protection systems (TPS); integrated vehicle health management (IVHM); quick turnaround operations; hypersonic air-breathing propulsion; and many others. NASA and Air Force planning are discussed next.

NASA Planning

Future NASA missions are planned to include human spaceflight and logistical support for the ISS, Earth and space science, astrophysics, and exploration of the solar system. In addition, NASA may seek to develop a new launch capability for very heavy lift to facilitate both crewed and uncrewed exploration of the solar system. Until recently NASA had been funding the development of an orbital space plane (OSP), which was to become operational before the end of this decade. Initial capabilities were intended to satisfy ISS crew rescue requirements. Crew transfer capability (Earth to LEO) was planned as a follow-on spiral development. The OSP was to be launched on an existing expendable vehicle modified to become crew-rated. It should be noted that since the OSP was a current vehicle development program and not a technology development program, it was not considered part of NAI. It is only mentioned here for the sake of completeness regarding future planned capability and for its potential to compete for limited NASA resources. However, the vision articulated by President George W. Bush on January 14,

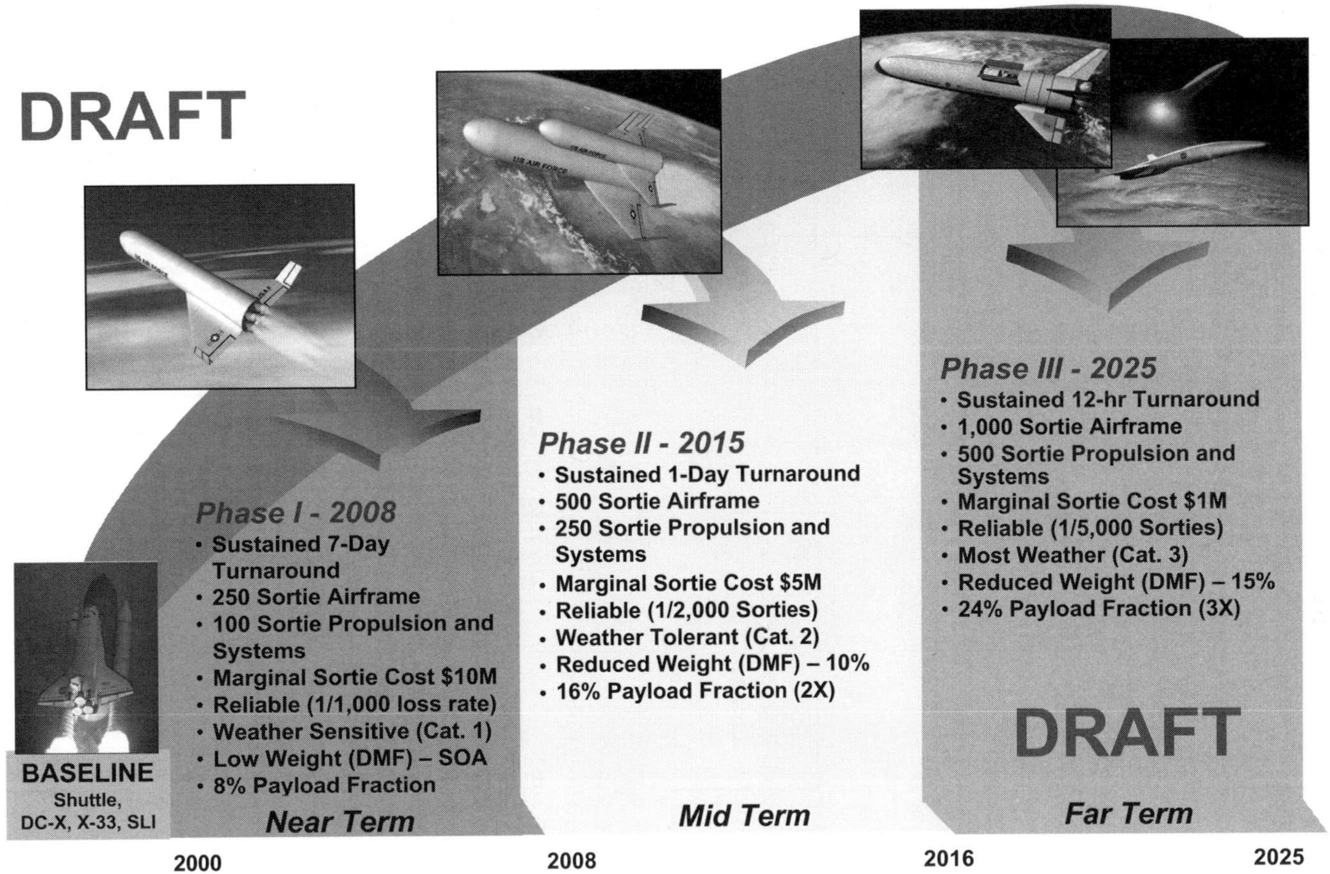

FIGURE 3-1 NAI phased approach to space access. Potential system payoffs and requirements. SOURCE: Sega, 2003.

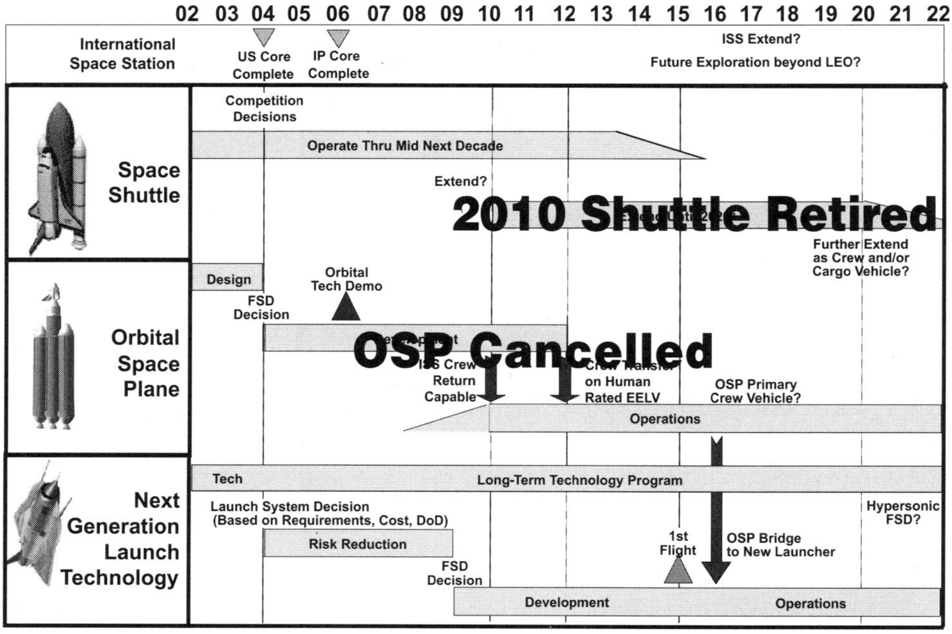

FIGURE 3-2 Integrated space transportation plan influenced by Presidential Vision announcement of January 14, 2004. SOURCE: Rogacki, 2003.

2004, called for the cancellation of the OSP in favor of a crew exploration vehicle (CEV) to be developed and tested by 2008.[1]

Like its predecessor, the CEV is initially planned to satisfy the ISS crew transport and eventually the ISS crew rescue missions. By any measure, the CEV development schedule is very aggressive. Compounding the obvious issues of funding and schedule is the apparent requirement (articulated in the Presidential Vision) to use this same vehicle to provide a crewed capability to destinations beyond low Earth orbit. It is assumed that the CEV will not fall under the NAI umbrella, but, like the OSP, it could compete for NAI resources within a tightly constrained NASA budget. This problem could be exacerbated if NASA finds that much of the CEV technology development is NASA-unique—an example would be technologies to support the beyond-Earth-orbit capabilities that would not be needed by the Air Force—and would therefore have to be funded at the expense of other common technologies sponsored by NAI. The President's Vision calls for retiring the space shuttles in 2010. After that, cargo up-mass and down-mass capability to the ISS will have to come from adapting evolved, expendable launch vehicles (EELVs) to the cargo mission, utilizing future European and Japanese cargo vehicles, or developing a new U.S. cargo capability (Williams, 2000). In parallel, NASA is developing next-generation launch technology (NGLT) to enable a decision on a rocket-based reusable launch system by 2008. Air-breathing hypersonic technology is not baselined for any near-term space access application but is under consideration for a long-term, crewed launch vehicle (Rogacki, 2003). Detailed plans and roadmaps for these new development programs (other than NGLT) were not available to the committee at the time of this writing. The top-level timelines for these activities are shown in Figure 3-2.

[1] See http://www.nasa.gov/pdf/54868main_bush_trans.pdf.

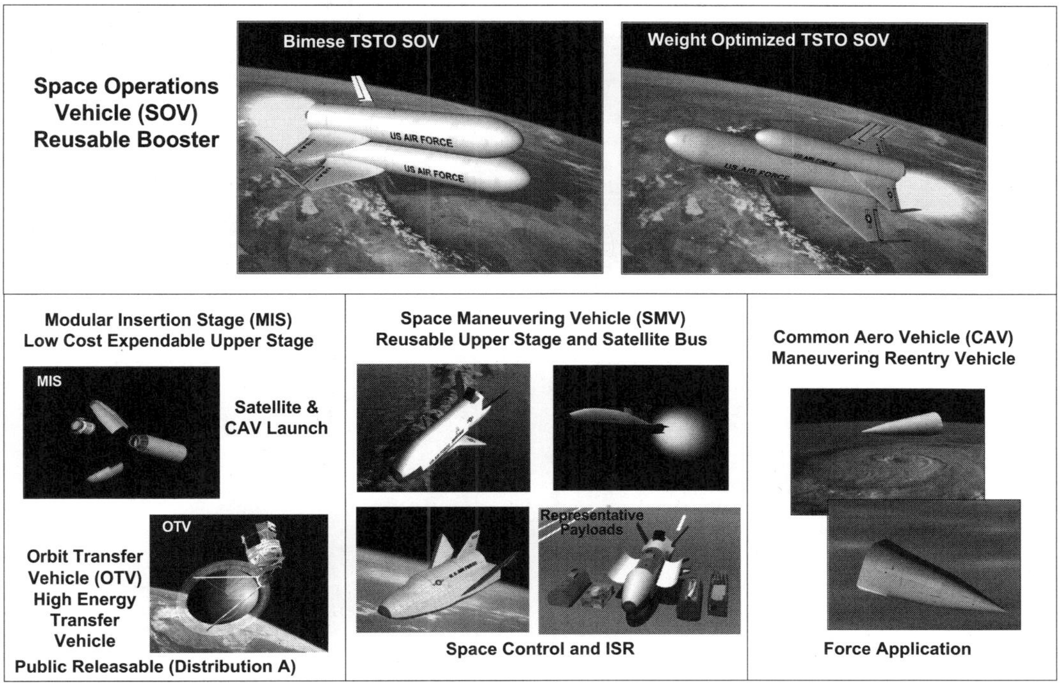

FIGURE 3-3 Military space plane system architecture. SOURCE: Koozin, 2003.

Air Force Planning

The Air Force's increasing emphasis on missions such as Prompt Global Strike (PGS), Operationally Responsive Spacelift (ORS), and Space Control (SC) may require both rapid-launch, expendable systems and reusable launch systems capable of turnaround times measured in hours rather than weeks or months. A concept for satisfying many of these requirements is shown in Figure 3-3. A potential access-to-space element of the architecture includes a two-stage-to-orbit (TSTO), reusable space operations vehicle (SOV) (booster) mated with either an expendable or a reusable upper stage. The associated development schedule for these vehicles is shown in Figure 3-4.

The Air Force may find a low-cost, time-responsive, expendable vehicle to be a cost-effective solution to meeting certain mission requirements. Upgrades to the current fleet of expendable vehicles may also prove beneficial. However, as demand for a responsive launch capability and the corresponding launch rates increase, reusable launch systems become more attractive, both operationally and economically. The propulsion system for both expendable and reusable systems will remain rocket-based in the near- to mid-term (2004-2018) owing to the immaturity of air-breathing engine technology. It is hoped that near-term demonstration vehicles (e.g., RASCAL and FALCON) will help to advance air-breathing engine technology and should provide data to help determine the feasibility of scaling these engines for medium to heavy lift vehicles.

Because air-breathing hypersonics technology is not baselined for any near-term space access concept but is being considered for longer-term designs (Rogacki, 2003), the committee's assessments in this chapter focus on rocket-based launch vehicles. The assessment of air-breathing-based launch vehicles is restricted to comments on the feasibility of the basic technology and on the data needed for deciding if air-breathing engines can be applied to launch vehicles for medium and heavy loads.

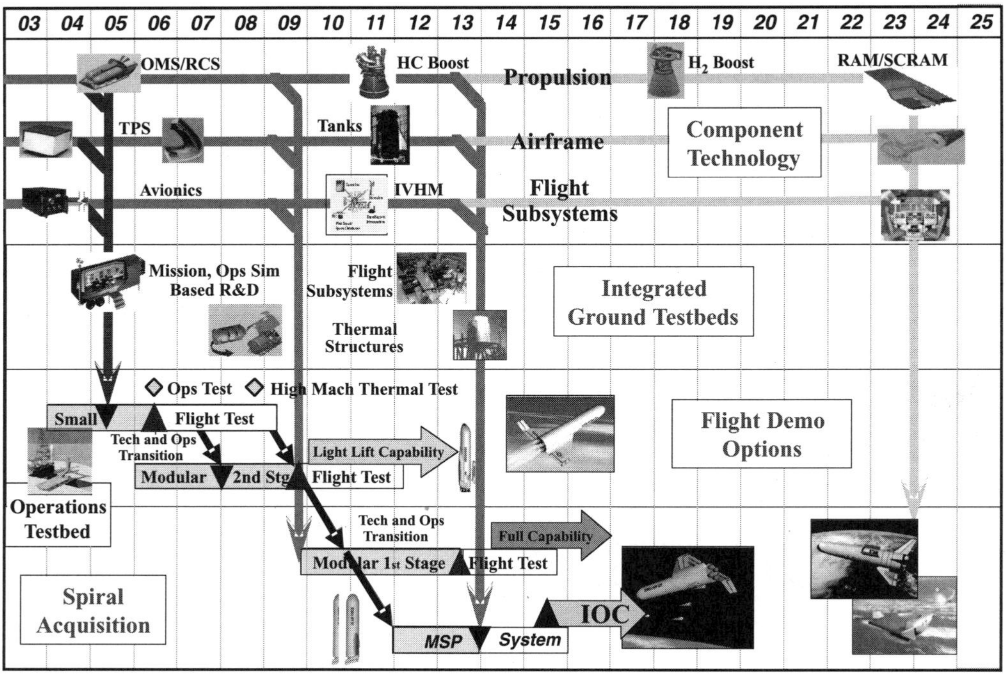

FIGURE 3-4 Military space plane roadmap. SOURCE: Sponable, 2003.

Finding 3-1. The United States will continue to rely on rocket-based systems for access to space for at least two more decades. The current expendable rocket families provide adequate lift capability for scheduled uncrewed launches to Earth orbit but will not support emerging military needs for rapid-rate, time-responsive launches. The two latest versions in the expendable rocket families utilize a new U.S. engine (RS-68) of undemonstrated reliability and an engine available only from a foreign supplier (RD-180.) The only reusable launch capability, NASA's Space Transportation System, is scheduled to be retired by the end of the decade. Owing to the loss of the shuttle Columbia and the resulting Presidential Vision announcement, the plans for crewed flights are undergoing extensive revision. That said, it is anticipated that human flight will be accommodated by the development of a CEV fitted atop a (yet-to-be-defined) booster. Compounding the obvious funding issues are the CEV multimission requirements, the aggressive schedule, and the likely desire to make the CEV compatible with both the Delta and Atlas families of expendables. Although not a part of NAI, the NASA-unique nature of the CEV and resulting demand on the NASA budget may affect NASA's ability to participate in NAI-sponsored technology development programs. The CEV booster might be further augmented and/or replaced in the next decade by a rocket-based, reusable launch system. Robust, reusable access to space will require a phased flight-test program to demonstrate reliability, affordability, and responsiveness relative to expendable systems. Air-breathing hypersonics technology is being considered for crewed and/or uncrewed launch systems in about two decades.

Recommendation 3-1. The Air Force and NASA need to strike a balance between vehicle development programs and the NAI-sponsored technology development programs. Both organizations should continue to invest in basic and applied research, technology development, demonstration, and implementation leading to new rocket-based capabilities. These investments should address both expendable vehicles and technology appropriate for deploying and operating reusable, rocket-based launch systems. Of near-term importance is the need to ensure the reliability and availability of the propulsion units used by the latest Delta and Atlas expendable boosters. Technologies developed under NAI sponsorship should be infused into the RS-68 under a component improvement program aimed at accelerating the near-term maturation of that engine. The question of assured availability of the Russian-built RD-180 used by the Atlas should be revisited in light of the potential increased demand following the loss of Columbia and the ensuing Presidential Vision announcement.

TECHNICAL FEASIBILITY IN NAI TIME FRAME

This section follows the taxonomy in the draft NAI plan for access-to-space S&T (DDR&E, 2002). It summarizes, in table form, the technical feasibility of each major technology area and its constituent technologies. The committee was hindered in its evaluation by the uncertainty surrounding budget projections. Consequently, the following subsections evaluate the various constituent technologies on the basis of their current maturity and their potential contribution to the overall NAI objectives of cost reduction, launch availability, and increased mission reliability and safety. An individual table is devoted to each NAI-designated vehicle system:

- Airframe
- Propulsion
- Flight subsystems
- Launch operations
- Mission operations
- Software

Each table further divides the vehicle system into major technology areas, which are in turn broken down into the constituent technologies defined (for the most part) by NAI. The tables provide the committee's assessment of (1) the current maturity of the constituent technologies, (2) the impact of the constituent technology on the major technology area, (3) the likelihood of achieving the desired results, and (4) the overall criticality of the major technology areas. Both the impact and criticality evaluations are based on the committee's estimate of their contribution to the overall NAI goals—economics, safety, reliability, responsiveness, performance, or benefit to the industrial base. Impact was described as negligible, little, moderate, significant, or extreme. Maturity levels, from 1 to 5, are not to be confused with TRLs; rather, they correspond roughly to the DoD S&T categories:

1. Basic research,
2. Applied research,
3. Advanced development,
4. Demonstration/validation, and
5. Engineering and manufacturing development.

According to the committee, constituent technologies with low maturity levels, high impact, and low likelihood of achievement should be afforded special attention if they lie within a major technology area with high criticality. Conversely, technologies with high maturity levels, low impact, and high likelihood of achievement should retain the current level of attention or less, depending on the criticality of the major technology area. The assessments in these tables should help to determine if appropriate attention is being paid to the development of technologies supporting NAI goals.

Finding 3-2. The committee assessed the technical feasibility of NAI technical objectives by examining the main technical challenges and evaluating the NAI response to those challenges. The committee's assessments are summarized in a set of tables corresponding to the NAI taxonomy.

Finding 3-3. The committee was hindered in its evaluation of financial feasibility by the uncertainty surrounding budget projections.

Recommendation 3-2. NAI should compare its current and projected resource allocations with the information contained in this report to assure appropriate attention is devoted to technologies with the best payoff in terms of costs, benefits, and risk.

Airframe

Depending on their configuration, the airframes of launch vehicles can serve several purposes, including thermal protection; enabling favorable lift-to-drag and thrust-to-drag ratios; housing and protection of onboard electrical equipment, pumps, feed systems, and engine systems; and fuel storage. The airframe is expected to be fully active in aerodynamic control of the vehicle. The external uncooled sections of the airframe may easily experience recovery temperatures in excess of 4000°F at Mach numbers of 8 and above (NRC, 1998). Compounding the mechanical and thermal loads on the airframe are the formation and movement of strong shock waves on the surface of the airframe; hot gas-solid (catalytic) chemical reactions on the surface; and unsteady boundary layers with flow separation. Several challenges must be overcome to operate the airframe effectively and efficiently in a complex aerothermodynamic environment. These challenges include extreme thermal, vibrational, and dynamic loading, and reliability, maintainability, and survivability, along with their impact on maintenance operations. Component and subsystem development leading to space launch capability faces challenges of accurate and efficient simulation and integration (for design), experimental testing (ground and flight), and, most critical, scalability.

Table 3-1 captures the primary technology issues for airframes supporting space launch. First, the committee found that none of the identified technology areas was fully matured. Second, it rated relative criticality as "high" in three of the four major technology areas and as "medium" in the fourth. Third, the thermal protection technologies and design and analysis tools were rated "low" or "medium" with respect to the likelihood of achieving their goals. Further development of the technologies may be costly and risky and will require a multidisciplinary research and development (R&D) framework to ensure system-level optimization. Ongoing and planned R&D activities in each of the three technology areas rated with "high" criticality are summarized below.

Thermal Protection

Three research areas with high-payoff potential have been identified and are summarized in the findings below.

Finding 3-4. The integrated high-payoff rocket propulsion technology (IHPRPT) program has a materials system component that addresses oxygen-compatible superalloys and metal matrix composites (MMC) for liquid oxygen turbopump housing (Brockmeyer, 2003).

Finding 3-5. The IHPRPT program has a three-component ceramic matrix composite (CMC)

TABLE 3-1 Status of Technologies for the Airframe System

NAI-Defined Technology Areas		Committee Evaluations			
		Constituent Technology Ratings			
Major Technology Area	Constituent Technology	Current Maturity Level (1-5)	Relative Impact on Major Tech Area (Negligible, Little, Moderate, Significant, Extreme)	Likelihood of Achieving in Time Frame (Current Funding Level) (Low, Medium, High)	Relative Criticality of Major Technology Area (Low, Medium, High)
Thermal protection	Materials	3	Extreme	Medium	High
	Leading edges	3	Extreme	Medium	
	Control surfaces	4	Significant	Medium	
	Acreage surfaces	3	Extreme	Low	
	Seals	2.5	Significant	Low	
	Hot structure	2	Significant	Low	
	Active/passive cooled CMC	2.5	Significant	Low	
Propellant tanks and feed systems	Reusable metallic	3	Significant	Medium	Medium
	Reusable cryo composites	1	Significant	Low	
	Insulation	4	Significant	High	
	Leak detection	4	Moderate	High	
Integrated structures	Isogrids	3	Moderate	Medium	High
	Highly integrated subsystems	3	Significant	Medium	
Design and analysis tools	Integrated design environment (modeling)	2	Extreme	Low	High
	Vehicle CFD/aerothermal	2	Extreme	Low	
	Structural design	3	Moderate	Medium	
	Cost and safety analyses	2	Moderate	Low	

materials initiative addressing thermal protection for bearings, nozzles, thrust chambers, gas generators, hot gas ducting, and turbomachinery (Brockmeyer, 2003).

Finding 3-6. The NAI and the NASA NGLT programs have identified critical and enabling airframe technologies for far-term development and use (McClinton, 2003).

Recommendation 3-3. The potential for high payoffs exists in each of these technology research areas, and NAI should continue to support them.

Integrated Structures

The National Aerospace Plane (NASP) and X-33 programs resulted in several advances in integrated structures. Much, however, remains to be mastered to ensure the reliability, safety, and short turnaround time of full-scale systems. Although integrated structures are one of the largest single contributors to improving vehicle mass fraction, very little experience exists in developing approaches and validating techniques for the engineering and manufacture of full-scale integrated structures.

Finding 3-7. The experience base of highly integrated structure, components, and subsystems is very shallow, and much time is required to develop new materials and integration techniques.

Recommendation 3-4. Research should concentrate on the development of (1) tools for the automated optimization of configuration, (2) design processes, (3) material characterization, and (4) testing to validate techniques and processes that enable highly integrated structures. NAI should recognize the time-intensive nature of this research and ensure that the effort is integrated into the overall NAI timeline.

Design and Analysis Tools

Model development is a component of the R&D program managed by NASA in its NGLT program. Design and analysis tools in support of two large areas of research remain underdeveloped. The first is the hypersonic reentry environment. These tools would be applicable for either a rocket-based or air-breathing reentry vehicle. The second area is the flow path environment for air-breathing hypersonic propulsion, both turbine and nonturbine. The need for robust and high-fidelity design tools and for ground-based testing to verify and validate these tools is well documented.

Finding 3-8. Some examples of numerical tools correlated to national ground testing facilities can be found. In general, numerical tools have not been validated by test data (Candler, 2003; Holden, 2003).

Recommendation 3-5. NAI should support programs to validate numerical tools using ground-based and/or airborne testing.

Summary

Lightweight reusable fuel tanks and highly integrated structures are significant contributors to increasing the overall system margin and the payload fraction of launch vehicles. Since the 1970s, NASA has been performing R&D on hypersonic airframe integration, with intermittent progress. The X-30 NASP exemplifies NASA's involvement in this area. Key airframe technology shortfalls were identified and risk assessments were made following NASP (McClinton, 2003). The NGLT value stream process—which is an interactive, multidisciplinary process to guide technology development—appears to be mature. Detailed work breakdown structures and various technology shortfalls are identified (McClinton, 2003). However, funding levels were not available to the committee. Generally, these programs have a goal of safe, affordable access to space while supporting DoD's recently elevated interest in Global Strike, Global Response, and Responsive Access to

Space. Some airframe technology metrics, technology readiness levels, technology gaps, and value streams have been established. A disciplined technique called goals, objectives, technical challenges, and approach (GOTChA) has been used to identify and map some technology development targets (DDR&E NAI, 2003). This technique was used to generate a state-of-the-art development plan. However, no detailed GOTChA was observed for the other 17 constituent technology areas listed in the second column of Table 3-1.

Recommendation 3-6. Congruent roadmap taxonomies with consistent metrics and TRL goals should be established between the NASA NGLT program and the NAI GOTChA results.

Recommendation 3-7. Project-level descriptions of airframe technology development approaches should be developed, integrated, and facilitated across government, industry, and university participants.

Propulsion

The state of development of rocket propulsion vis-à-vis that of air-breathing propulsion points to the exclusive use of rocket propulsion in the near and medium term (2003-2015). By 2015, air-breathing propulsion may have advanced enough to put small assets (<500 kg) into LEO as the first stage of a system with a rocket-propelled second stage. Heavy lift (2,000-20,000 kg) with air-breathing propulsion will only mature in the far term (2025 and beyond). Turbine engines must be significantly scaled up in thrust and in Mach number capability if they are to support heavy-lift horizontal takeoff. Also, airframes must be developed with adequate support systems—TPS, structures, materials, controls, and so forth—to be viable. (Development of these support systems would also benefit rocket propulsion vehicles but might go beyond what is required by rocket systems.) The development work to enable a reusable, heavy-lift, air-breathing propulsion vehicle will be extensive, costly, and require basic research in many areas. Therefore, in the near to medium term, reusable, rocket-propelled, heavy-lift vehicles that take off vertically and land horizontally appear to be the best approach to meeting the access-to-space requirement. The assessment in Table 3-2 considers rocket-propelled vehicles and follows the taxonomy specified by the NAI program (DDR&E, 2002).

It should be noted that the NASA NGLT program not only targets reduced cost, faster turnaround time, and more reuse than the current shuttle capability but also specifies a lift capability of 30 to 50 metric tons. Only NASA has embraced this requirement. The cost reduction that might be realized from increased vehicle utilization will not materialize if NASA remains the only entity requiring this high-lift capability.

Finding 3-9. Only NASA specifies a requirement for a reusable, heavy-lift (~40 metric tons) space launch vehicle.

Recommendation 3-8. NASA should reevaluate its heavy-lift requirement. A trade study between high lift capacity and alternative approaches to meeting mission needs utilizing a common vehicle should be conducted, especially in light of the recently demonstrated ability to construct a large space structure, the ISS, in orbit. Adjusting the NASA lift requirement to match that of other users (e.g., DoD) would increase utilization and save billions of dollars for the development of a unique, low-flight-rate vehicle.

A TSTO approach with a liquid oxygen (LOx)/HC first stage and a LOx/liquid hydrogen (LH_2) second stage appears to be the configuration baseline on which propulsion improvements are to be made. The goals are to reduce the failure rate, increase specific impulse (Isp), reduce hardware and

TABLE 3-2 Status of Technologies for the Propulsion System

NAI-Defined Technology Areas		Committee Evaluations			
		Constituent Technology Ratings			
Major Technology Area	Constituent Technology	Current Maturity Level (1-5)	Relative Impact on Major Tech Area (Negligible, Little, Moderate, Significant, Extreme)	Likelihood of Achieving in Time Frame (Current Funding Level) (Low, Medium, High)	Relative Criticality of Major Technology Area (Low, Medium, High)
Propellants (oxidizer/fuel)	LOx/HC (O_2 rich)	3/5[a]	Significant	Low	Medium
	LOx/LH_2	5	Significant	Medium	
	H_2O_2/HC	2	Little	Low	
	High-energy green fuels	1	Little	Low	
Propellant management devices	Turbine pumps	3	Extreme	Low	High
	Engine lines	3	Extreme	Low	
	Engine ducts	3	Extreme	Low	
	Engine valves	3	Extreme	Low	
	Cryo level sensors	4	Significant	Medium	
Combustion and energy conversion devices	Chambers	3	Extreme	Low	High
	Nozzles	3	Extreme	Low	
	Injectors	3	Extreme	Low	
	Gas generators	3	Extreme	Low	
	Preburners	3	Extreme	Low	
Controls	Sensors	4	Moderate	Medium	Medium
	Health management	2	Significant	Low	
	Software	4	Significant	High	
	Engine controls	4	Significant	Medium	
Materials	O_2-rich compatible	3	Significant	Low	High
	High temperature	2	Significant	Low	

[a]3 for U.S. propellant; 5 for foreign propellant.

support costs, increase thrust/weight, and increase mean time between removals (reusable life) (DDR&E, 2002). The committee's assessment of the technologies contributing to propulsion is presented in Table 3-2.

The Integrated High Payoff Rocket Propulsion Technology (IHPRPT) program has made progress in engine design, materials, and performance in support of the access-to-space goals—for example, the Rocketdyne LOx/LH_2 engines, RS-68 and RS-83, and its LOx/RP engine, RS-84. In particular, their integrated powerhead demonstrations (IPDs) have shown significant progress toward longer operating life in an oxygen-rich environment, increased reuse, and reduced cost for LOx/LH_2 engines. However, funding limitations have not allowed any investment in LOx/RP engine advances.

Propellants

Both LOx/HC and LOx/LH_2 engines are the basic propulsion engines for today's space access rockets. They are proven and reliable for single-time use (space shuttle main engines are "reusable," albeit after inspection and refurbishment). To increase the Isp, O_2-rich mixtures can be used. However, the temperatures and associated chemical environments created by an O_2-rich environment degrade existing engine materials (the exception is the Russian-built RD-180, which still has not been built in the United States). Therefore, engine material development is required to realize the performance increase that can be obtained from O_2-rich mixture ratios.

The Air Force is working on an H_2O_2/HC engine for payloads such as the space maneuvering vehicle (SMV), the modular insertion stage (MIS), and the common aero vehicle (CAV) (Kastenholz, 2003). This fuel combination requires material development to limit its reaction with engine materials. This work is ongoing but severely limited by funding. A decision needs to be made on whether to continue this development or to rely on improving engine performance using LOx/HC and LOx/LH_2.

High-energy, green fuels (space propellants that are environment-friendly and nontoxic) are a long-term goal of the NAI access-to-space pillar. No work was reviewed on this subject. Fuels R&D work should be funded if their use is a goal of the program.

Finding 3-10. LOx/LH_2 and LOx/HC engines run with oxygen-rich fuel mixtures can achieve better engine performance (Isp).

Propellant Management Devices

Reduced cost, chemical compatibility, higher operating temperature, lighter weight, and longer life (reuse) is needed for all propellant management devices. This can be achieved through design and materials advances. Since design and materials are coupled and funding for materials research is very limited, progress is slow. Certainly advances can and have been made through design alone, but to realize the required engine performance, sufficient funding must be allotted to advance materials as well.

The exception to this conclusion may be cryo-level sensors, which are currently in use. Their reliability can be improved, but that does not appear to entail a large development effort.

Finding 3-11. Advances in materials and design can reduce cost, increase chemical compatibility, enable higher operating temperatures, reduce weight, and increase the operating life of propellant management systems.

Combustion and Energy Conversion Devices

NASA and industry are addressing the development of engines using fuel-rich staged combustion (FRSC) and oxidizer-rich staged combustion (ORSC). These programs include advances in combustion devices, turbomachinery, engine systems, and model development (Boeing, 2001). This R&D initiative includes gas seal improvement, elimination of welds, hydrostatic bearings, and simplifying design by reducing part count. This suite of technology improvements should be available for scale-up in the 2010 time frame.

As with propellant management devices, reduced cost, chemical compatibility, higher operating temperatures, lighter weight, and longer operating life (reuse) are required of these devices. In addition, better mixing of the fuel and oxidizer is needed at the proper mixture ratios and flow rates. Significant additional funding is needed for materials development, design, and device testing to realize the engine performance goals on the planned schedule.

Finding 3-12. Advances in materials and designs of combustion and energy conversion devices are needed to reduce cost, increase chemical compatibility, increase operating temperature, reduce weight, and increase operating life.

Controls

Of all the engine components, controls (except for health management systems) require the least development. Increased reliability is the main goal.

On the other hand, the health management system, a key enabler of multiuse engines, is far from fully developed. New sensing methods and prediction algorithms are needed to predict the effects of crack propagation, vibration, mixture ratios, temperatures, and so forth on material health and longevity. Significant development is required in this area, coupled with reliable, low-maintenance designs that afford longer life (reuse) and shorter turnaround times.

Finding 3-13. Engine health management systems have the potential to make significant contributions to the goal of aircraftlike engine operations.

Recommendation 3-9. NAI should define and implement a plan to develop and test an engine health management system that, coupled with reliable, low-maintenance engine designs, will realize the goal of reuse with short turnaround times.

Materials

Advanced materials are the primary key to obtaining desired engine performance. Research, development, and testing of materials (and devices fabricated from these materials) are needed. Advances in ceramic matrix composites have been made, but further advances in these and other materials are necessary to make use of higher-energy fuels in reduced weight engines that function for 100 to 500 launches with minimum maintenance. Funding for materials development is too little to realize long-term NAI goals.

Finding 3-14. Advances in engine system materials are required to realize enhanced engine performance with multiple reuse and fast turnaround times.

Recommendation 3-10. NAI should define and implement a strong materials development and test program that enables the use of oxygen-rich fuel mixtures and results in engine systems with increased margins, long operating life, and reuse with short turnaround.

Summary

Propulsion advancement is a key driver for increasing the payload fraction and reducing cost and turnaround time for launch vehicles. Funding and programmatic emphasis need to be devoted to this area to increase engine operating margins and reliability and thereby realize NAI program goals.

Flight Subsystems

Integrated flight subsystems are a key enabler of increased vehicle performance and operability. Future visions of aerospace vehicles in both the civilian and military sectors are based on plans that require wide-ranging evolutionary changes. In accessing space, the significant difference between future requirements for commercial/civil vehicles and those for the military space plane system is the military requirement for a responsive space lift capability (Sega, 2003). Flight subsystems are impacted strongly by the specific technology goals identified to enable this capability. For example, future military requirements will include aircraftlike operability—that is, activation for service within days and launch within hours; greatly expanded maneuverability over a larger operating regime; significantly enhanced reliability; and much lower operational costs—which will bring revolutionary changes in vehicle reusability, self-management, adaptability, intelligent systems, and modularization. Thus, for integrated flight systems, general access-to-space goals that include reliability, decreased cost, reduced weight, and increased operability will necessitate critical developments of flight subsystems.

Table 3-3 summarizes the committee's evaluation of the flight subsystems.

Power

Generation. Current power generation systems (hydraulic, electric, and pneumatic) are heavy and inefficient. Several layers of redundancy, necessitated by low reliability, contribute to weight. These systems also require excessive maintenance between flights and are not designed for automated maintenance. Also, some power generation systems use toxic fuels.

NAI Phase I efforts already focus directly on heavy and unreliable hydraulic systems. Phase II will give special attention to increasing the robustness and decreasing the weight of power systems, with a focus on flightworthy hardware in preparation for an eventual Phase II flight test. The committee believes the power system should be a high priority. DoD has already identified it as a key enabling technology (DDR&E, 2002).

Storage. Energy storage is a part of the power system that is also receiving significant attention under Phase I. By Phase II, a less complex system is envisioned—one that includes advanced fuel cells and conformal batteries with decreased maintenance. Phase III is expected to include innovations in battery chemistry that will further increase life and reduce material toxicity. With such a high priority given to this constituent technology so early in the plan and in light of known development objectives, success is likely.

Distribution. Phase I advances in power distribution correlate with those in power generation. Phase I is also introducing a distributed power architecture and, potentially, fault-tolerant photonic

TABLE 3-3 Status of Technologies for Flight Subsystems

NAI-Defined Technology Areas		Committee Evaluations			
		Constituent Technology Ratings			
Major Technology Area	Constituent Technology	Current Maturity Level (1-5)	Relative Impact on Major Tech Area (Negligible, Little, Moderate, Significant, Extreme)	Likelihood of Achieving in Time Frame (Current Funding Level) (Low, Medium, High)	Relative Criticality of Major Technology Area (Low, Medium, High)
Power	Generation	4	Extreme	High	High
	Storage	3.5	Extreme	High	
	Distribution	4	Extreme	Medium	
	Photonics	1	Moderate	Low	
Actuation	Photonics	2	N/A	Medium	Medium
Vehicle management	Adaptive GNC	2	Moderate	Low	High
	Sensors	3	Significant	Medium	
	Hardware obsolescence planning	3	Significant	Medium	
Thermal cooling	Materials	4	N/A	High	Low
Vehicle health	Sensors	3	Extreme	Medium	Medium
	Prognostics	2	Significant	Low	
Autonomous control of flight mechanics		2	N/A	Low	High
OMS/RCS	Propellants	3	Significant	Medium	Medium
	Ignition systems	3	Significant	Medium	

NOTE: GNC, guidance, navigation, and control; OMS, orbital maneuvering systems; RCS, reaction control system.

controls. Power distribution, particularly with photonic control, is not as well developed as power generation and storage.

Photonics. In Phase I, which is relatively short, NASA and AFRL will attempt to exploit current subsystem technologies. Reliable photonics for vehicle management is a 6.2 research effort, so photonics is not expected to be available until at least Phase II. Ground demonstrations are designed to validate the reliability of flight-critical optic sensors for optically controlled power switching. Not only must the key technologies be matured, but photonic optics must be integrated with the vehicle management system. Flight test will ultimately be dependent on the critical reliability of the photonic vehicle management system (VMS). NAI is relying heavily on the successful development of photonic technologies (DDR&E, 2002).

Finding 3-15. In the near term, the control of power generation, storage, and distribution depends on currently available technologies. Future capabilities will rely on the integration of photonically controlled power generation, distribution, management, and actuation systems coupled with the photonic VMS.

Actuation

Consistent with the evolution of power systems, Phase I is beginning a shift from hydraulic to electric actuation systems. Integrating power and actuation should reduce risk in the overall system. Demonstration of increased reliability in actuating components (including aero surfaces and thrust vector control) is also expected be accomplished in Phase I. Fiber-optic components and sensors for control will be introduced in Phase II. Investments are expected in technologies to improve motor efficiencies. Phase III goals target efficient, long-life components.

Vehicle Management

As noted in the draft access-to-space S&T plan (DDR&E, 2002), Phase I focuses on tailoring state-of-the-art aircraft technology and adapting it for reusable lift vehicles and space environments. Ground demonstrations are used to assess system-level components. As such, a number of programs in the vehicle management system (VMS), integrated vehicle health management (IVHM), and guidance, navigation, and control (GNC) will be brought together and validated in a simulation environment. This is expected to establish GNC technologies that will lead to fully autonomous systems. The initial phase relies on off-the-shelf components. However, 6.2 research in this same time frame includes development of a reliable photonic vehicle management system for data acquisition, dissemination, computation, and control. By Phase II, the vehicle management system is to be integrated with a vehicle health monitoring system having sensors, intelligent algorithms, and software for prognostics and diagnostics at a subsystem level. As components degrade, the VMS/IVHM system will identify performance penalties and implement adjustments. This is a revolutionary development program and, again, will rely on a photonically integrated architecture with validation using high-fidelity, ground-based simulation. A photonic VMS will also bring decreased weight and improved reliability. "Building" the VMS system involves information technology and advances in computational capability, both of which have received a lot of attention and achieved incredible progress in a wide range of commercial activities.

Finding 3-16. NAI applications will benefit (technically and, possibly, in terms of reduced costs) by exploiting computational (both hardware and software) developments in the commercial sector.

Recommendation 3-11. NAI should assess its current vehicle management system development program to ensure that technology available in the civilian sector is leveraged as much as possible.

Advanced GNC is required for integration with the VMS system. However, Phase I expectations are more modest: a fault-tolerant system and validation of tools for GNC design. Investments in Phase II are planned to support the development of intelligent and adaptive VMS systems, including 6.2 research in autonomous and adaptive GNC. Ground demonstrations and simulation will integrate the VMS with advanced flight GNC and will incorporate photonic VMS. Correspondingly, adaptive GNC implies real-time mission planning and replanning. Ground-based simulation and flight testing will validate the integrated adaptive system. This integrated, intelligent, adaptive photonic VMS-GNC technology is universal across the three NAI pillars. Any delay in the research and/or technology development timeline will have impacts across all three pillars.

Reliable optic sensors for control of the VMS system are flight-critical components. They must be fully mature and reliable if they are to be integrated into the VMS.

The hardware developed for future flight systems is expected to have longer life, with less maintenance and a longer time between failures. A modular framework supports these requirements and allows component elements to be updated as needed. A response to hardware obsolescence should be incorporated into the system/component development plan.

Finding 3-17. Integrated, intelligent, adaptive photonic VMS-GNC technology is universal across the three pillars of NAI. Any delay in the research and/or technology development timeline will have impacts across the entire NAI program. These systems will in large part be validated for flight using ground-based simulation tools.

Recommendation 3-12. The development timelines of the VMS-GNC program should be evaluated to ensure that the technology is ready whenever it is needed to achieve an integrated system and flight test within a decade.

Recommendation 3-13. Programs for vehicle simulation tool development should be periodically evaluated to ensure that they keep pace with the hardware and component development programs.

Thermal Cooling

Thermal cooling for local and distributed power systems is noted as a critical R&D area in the draft NAI plan for access to space (DDR&E, 2002). Thermal control is required to maintain reliability of high-temperature components including the electronics. By Phase III, a matrixed thermal management architecture is anticipated with high-performance active/passive thermal management. Although thermal cooling is a significant issue, R&D in other technology areas is more critical. Thus, in the context of flight subsystems, thermal cooling technologies can be effectively leveraged from other R&D programs.

Vehicle Health

Similar to the engine health monitoring discussed in the section on propulsion, a vehicle subsystem health management system to continuously monitor, diagnose, and prognosticate the vehicle status will be critical to improving vehicle availability. Initially, in Phase I, the ground demonstration for VMS and IVHM will integrate critical GNC technologies with flight information, with a focus on the robustness and reliability of the health monitoring systems. Ground demonstrations will then incorporate the integrated IVHM system in preparation for flight test. The IVHM system is designed to monitor the health of the power, actuation, and other subsystem components and then integrate the information into the VMS. However, sensor development is necessary to meet the requirements of the VMS/IVHM system.

Prognostics capability also relies on component sensors (hardware), plus intelligent and adaptive algorithms (software). Once implemented, the VMS/IVHM system must meet reliability and performance specifications.

Flight Mechanics and Autonomous Control

Flight mechanics, encompassing the control, navigation, and guidance of vehicles, is noted by NAI as a critical R&D area. It impacts operability and, to some extent, reliability as well, particularly with expanded flight regimes. Planned ground tests in Phase I include the modeling and simulation of autonomous flight mechanics capabilities in a virtual environment. This is also an area designated for 6.2 research in Phase II. Flight mechanics modeling is also critical to the development of the VMS/IVHM system—for example, in development of the prognostics capability. Responsive autonomous mission planning and replanning imply flight mechanics as part of the advanced, adaptive GNC system. By the Phase II flight test, a real-time autonomous/adaptive GNC system that relies solely on vehicle information will incorporate flight mechanics algorithms. This capability will include abort contingencies and real-time response. A significantly advanced software validation and verification program must also be in place as early as possible. By Phase III, a sophisticated integrated system is planned to incorporate nanotechnology for computing.

Orbital Maneuvering System/Reaction Control System

The elimination of toxic propellants is a high priority. A shift toward more electric-actuation systems also reduces the operational burden by eliminating costly and potentially unsafe maintenance requirements. The evaluation of ignition systems is based on an overall integrated OMS/RCS system actuation, control, and reliability assessment.

Launch Operations

In the launch operations arena attention has been concentrated on work site safety, environmental compatibility, and reductions in vehicle-unique operations, with a corresponding reduction in the overall number of operations and their labor intensiveness. "Aircraftlike operations" has become the mantra of those hoping to instill a new, cost-effective method of performing launch operations. By almost every metric, the process of planning, preparing, and launching a rocket is vastly more expensive and cumbersome than similar activities for even the most notoriously complex and archaic aircraft.

Launch vehicles operate in a much more demanding environment than that in which typical aircraft operate, and they must achieve a much higher level of performance. By any measure (e.g., weight, cost, payload fraction), the threshold of adequate performance forces the designer to narrow the gap between the acceptable operating point and the ultimate physical limit. It is for this reason that launch systems are designed with much smaller operating margins. This lack of margin in launch vehicles (as compared with more forgiving aircraft) is recognized as the fundamental reason for the great disparity in operating costs. Consequently, it is advances in the basic design and engineering of each vehicle that will eventually result in the biggest savings in launch operations. Adding robustness to the vehicle will allow mission planning, ground processing, and flight operations to more closely achieve the efficiencies that new technologies will enable. Because some of the biggest launch process drivers are unique to specific launch vehicles, it is difficult to imagine a complete set of generic areas where advances (cost reductions) can be made independent of the vehicle configuration. Indeed, the processes and technology thrusts must be traceable and scalable to the envisioned vehicle.

It seems that NAI has recognized this fact. It has embarked on a series of technology investigations to enhance the operating tempo and reduce the costs of traditional launch processes.

TABLE 3-4 Phase I and Phase II Technology Goals for Various Launch Activities

Activity	Baseline (STS)	Phase I Goals	Phase II Goals
Propellant management	3 hours	2 hours	1 hour
System assembly	~4 months	24 hours	4 hours
Launch pad operations	2-3 weeks	24 hours	4 hours
System refurbishment	100,000 man-hours	7,500 man-hours	1,200 man-hours
Mission operations	100s of people	30 people	15 people
Range reconfiguration	48 hours	24 hours	12 hours

SOURCE: DDR&E, 2002.

> The overall ATS [access to space] goals are 7-day turnaround time and $10M marginal sortie cost for Phase I and 1-day turnaround time and $5M marginal sortie cost for Phase II. The thrust areas include propellant management, system refurbishment, system assembly, launch and pad operations, mission operations and range operations. Reference data, based on information obtained from the NASA Kennedy Space Center (KSC) for the space shuttle system, was used as baseline capabilities in the six areas for a reusable launch system. The NAI ATS Operations Technology portfolio only addresses improvement from operations technology enhancements and allocates process improvements to be investigated during system design & development. (DDR&E, 2002)

Top-level goals are depicted in Table 3-4. Additional Phase I goals copied from the draft plan require the launch operations to be accomplished in 24 hours. Six hours are allotted for damage assessment and an unspecified number of hours for refurbishment. Operations identified as labor-intensive include postflight inspection procedures, high-maintenance acoustic suppression systems, elaborate exhaust management systems, and umbilical inspection/refurbishment/checkout. Approaches to meeting these goals include durable, energy-absorbing materials, nonpyrotechnic hold-downs, fly-off disconnects, and intelligent sensor and inspection systems (DDR&E, 2002).

Another Phase I goal is to accomplish mission operations with no more than 30 people. NAI says that the approaches to overcoming the challenges are advanced control and monitoring systems, enhanced decision models, advanced weather instrumentation, rapid mission planning, simulation and certification, and operations control center simulation.

Several programs are under way to explore methods and procedures for reaching these goals. Objectives defined by AFRL for the integrated high-payoff rocket propulsion technology (IHPRPT) program are to develop affordable technologies for revolutionary, reusable and/or rapid response military global reach capability; sustainable strategic missiles; long life or increased maneuverability; and high-performance tactical missile capability. IHPRPT includes specific goals for launch operations in 2015, 2020, and 2025, with reductions in operations costs from 15 to 45 percent for expendable launchers and from 2- to 20-fold for reusables (DeGeorge, 2003). DARPA's RASCAL program is attempting to demonstrate a launch cost of $750,000 per mission on a recurring basis (~$2,000 per pound to orbit) (Singer, 2003). NASA's NGLT X-43 demonstrations are primarily engine focused but are also proposed for enabling durable, intelligent TPS, an all-electric launch system with health management, and a space-based launch tracking system (Lyles, 2003).

The NAI programs and plans for launch operations are assessed in Table 3-5. The evaluation of all but one of the technology areas can be applied to either a vertical or a horizontal takeoff vehicle. The evaluation of the Automated Ops technology area is based on a vertical launch vehicle. It is assumed that achieving or approaching these goals would result in reducing marginal sortie costs. Reusable launch vehicle (RLV) marginal sortie costs are a legitimate target of the NAI ground operations research. It is the goal of NAI to reduce the marginal sortie cost for a reusable launch

TABLE 3-5 Status of Technologies for Launch Pad Operations

NAI-Defined Technology Areas		Committee Evaluations			
		Constituent Technology Ratings			
Major Technology Area	Constituent Technology	Current Maturity Level (1-5)	Relative Impact on Major Tech Area (Negligible, Little, Moderate, Significant, Extreme)	Likelihood of Achieving in Time Frame (Current Funding Level) (Low, Medium, High)	Relative Criticality of Major Technology Area (Low, Medium, High)
Weather prediction	Winds aloft	4	Little	High	Medium
	Lightning prediction and control	2	Significant	Low	
Automated ops	Umbilical	3	Significant	High	High
	Hypergolic fuels	2	Significant	Medium	
	System calibrations	4	Moderate	High	
	Ground and range ops design and analysis tools	4	Moderate	High	
Integrated range network architecture	Onboard range safety	4	Extreme	High	Medium
	Responsive reconfiguration	3	Moderate	Medium	
Intelligent inspection	TPS acreage	1	Extreme	Low	High
	Health management	2	Significant	Low	
Security unauthorized penetration	Physical		N/A		Low
	Electronic		N/A		

system from approximately $300 million per sortie to $10 million per sortie (DDR&E, 2002). A majority of the reduction comes from a reduction of manpower for supporting flight and ground operations and the elimination of expendable elements and hazardous operations.

Finding 3-18. NAI defines the RLV marginal sortie cost to include salaries for all manpower during flight operation and ground turnaround operations, range costs, and propellant and other consumable costs. Although marginal costs are an important metric for the cost effectiveness of a system, marginal costs alone do not provide adequate insight into the cost of ownership. Total life-cycle costs combine marginal costs with the cost of money and the cost of acquisition, providing greater appreciation of system affordability. The attempt to reduce ground operation costs through the development of highly automated systems could require sizable expenditures. The cost effectiveness of such an investment is a strong function of flight rate and may not be justified when considered in the light of total life-cycle costs.

Recommendation 3-14. NAI should evaluate all potential cost-saving ground operations technologies against a cost model that includes nonrecurring acquisition costs and the cost of money.

Recommendation 3-15. NAI should evaluate the anticipated advantages and disadvantages of both reusable and expendable launch systems against a total cost model that includes nonrecurring acquisition costs, the cost of money, and other operational advantages.

NAI has planned an Operations Technology Roadmap encompassing six broad technology areas that are considered traditional cost drivers. Two ground demos and one flight demo are planned through FY 2014. Appendix D provides a brief summary of currently planned demonstrations. Significant results from system-level ground demonstrations that would be applicable to a wide range of potential vehicle configurations are unlikely. Although individual technology demos are useful, system-level ground demonstrations are costly. Going through the effort to combine individual technologies into a system-level demo might not be a cost-effective use of resources this far in advance of the known operational configuration.

Finding 3-19. It is difficult to show launch operations traceability and scalability without a further definition of the eventual vehicle configuration.

Finding 3-20. It is likely that a favorable cost/benefit ratio can be achieved by designing robustness into the vehicle.

Finding 3-21. The committee was not presented with evidence to validate the vehicle touch labor, marginal sortie costs, and turnaround time goals. It is not clear that achieving these goals will result in the most cost effective and operationally relevant launch capability, especially when considering the nonrecurring costs.

Finding 3-22. Traceability from ground demonstration to operational system is difficult without a clear understanding of the operational vehicle's final configuration.

Recommendation 3-16. NAI should evaluate the cost/benefit ratio of the two ground demonstrations. Consider concentrating resources on the flight demo.

X-42

The X-42 is envisioned to be a flying testbed for validating selected technologies developed since the late 1980s. Of the numerous test programs initiated over the last two decades, only one (the DC-X/A) has made it to flight status. Consequently, a significant investment in technology has gone untested in flight. NAI plans to correct this situation using the X-42 flight test vehicle.

In addition to being an environment simulator for airframe and flight subsystems technologies, NAI advertises that the X-42 will be used to evaluate improved technologies and approaches for fighterlike operability in a military space plane (MSP) type system. If designed properly, it can also be used as a payload integration testbed, enabling flight testing of the next generation of air-breathing propulsion technologies.

Flying demonstrators are ideally very traceable and scalable to a follow-on operational system. As envisioned today, X-42 would be scalable to the second stage of a TSTO space plane. Scalability ensures that component technologies will be developed and flight tested in environments relevant to the operational end system. NAI believes that the subscale nature of the X-42 and its ability to launch from the ground greatly simplify the complexity of the system—that is, subscale existing engines are adequate, flight lasts only 10-20 minutes, reaction control system (RCS) usage is limited, there are few or no auxiliary power units (APUs) and no need for on-orbit subsystems, and so on (DDR&E, 2002). It is reasonable to expect that a ground-launched test vehicle will simplify the effort as compared with an air-launched vehicle by eliminating the carrier aircraft restrictions, interface complications, and aerodynamic interference issues. Unfortunately, past experience has shown that the subscale nature of the X-42 will bring very little simplification to the vehicle. In fact, it will most likely introduce an entire new set of complexities related to scaling (e.g., thrust to weight differences). Furthermore, hanging a test engine on a test vehicle is contrary to standard and prudent flight test practice.

Finding 3-23. A flight demonstration is long overdue. However, X-42 (and potentially the X-43 series) test objectives appear to be oversubscribed.

Recommendation 3-17. NAI should limit the claimed X-42 objectives to a more reasonable number. Likewise, if the X-43 series (discussed previously) includes enabling durable, intelligent TPS, an all-electric launch system with health management, and a space-based launch tracking system, as advertised by NASA (Lyles, 2003), then it, too, will be overburdened. Too many objectives will dilute the effectiveness of each objective, increase overall risk, and increase the probability of failure.

Finding 3-24. NAI has inherited a plethora of planned ground and flight demonstrations from all three military services and from NASA and DARPA. These demonstrations address various aspects of the technologies necessary to advance the NAI access-to-space pillar.

Recommendation 3-18. NAI should strive to ensure that all these demonstrations are clearly defined, make significant contributions to the engineering and scientific knowledge base, and are adequately funded to achieve the desired objectives.

Weather Prediction

Weather prediction and monitoring play a big role in safely launching and servicing spacecraft today. Upper-level winds, lightning, surface winds, and ceiling all play havoc on launch and landing days. In addition, military operations are going to demand systems capable of flying in more adverse weather. Therefore, the ability to predict weather conditions with great accuracy in unstable atmospheric situations will be required. These conditions will need to be known during vehicle ascent and also during descent from orbit. Some of these obstacles will be partially addressed by new vehicle design requirements (near-all-weather capability, adaptive avionics suites, etc.), but there will always be a need for predicting weather and induced lightning that could jeopardize the safety of the mission. Everyday work is also impacted by local weather conditions when lightning storms move into the launch area and all outdoor activity is curtailed. All of these demands point to the need for more accurate forecasting to be available to operators and mission planners in real time. The criticality of weather prediction has been rated at "medium" due to the potential mitigating effects of other operational techniques (such as utilizing multiple launch sites or inland launch sites and pad protection facilities.)

Automated Operations

Specific developments in this area tend to be vehicle-architecture-specific. For example, number of stages, stacking orientation, ground/air/rail launch, and other factors all influence the degree of automation and the potential to automate. The NAI plan proposes to examine emerging system architectures for ground-handling requirements and approaches, including automated mating and assembly, component sensing and locating, and rapid ground power connections.

Umbilicals containing propellant, electrical power/signals, and cooling fluid connections require a significant amount of attention to maintain in reliable states. NAI intends to develop smart umbilicals to drive down the recurring cost and support high-tempo operations. These devices will automatically align themselves to the correct mating position and then perform a built-in test function. Nonpyrotechnic release technologies will be incorporated into the umbilicals to avoid the serial processing generally required when handling hazardous materials. Research into the automated handling of hypergolic fuels will also be initiated. The criticality value assigned to system calibrations is due mainly to the need to calibrate flight safety-critical landing aids. The overall criticality of automated operations has been rated "high" due to its significant contribution to manpower reduction and safety enhancements.

Integrated Range Network Architecture

Current range safety tracking and communication systems at the Eastern Test Range are based on 1950s technology. These systems, located near the launch site and spread many miles downrange, are very expensive to operate and maintain. Another hindrance is their inability to handle more than one launch vehicle in any 48-hour period. A space-based range architecture would provide a flexible network of tracking and communication links, enabling global launch operations. Furthermore, the only precision landing aid available at the Eastern Range is an antiquated microwave landing system used by the shuttle. Future RLVs will most likely incorporate a differential Global Positioning System (GPS) system owing to its lower costs, increased reliability, and growing acceptance as a global standard precision landing aid. NAI is planning improvements to ground sensors and instrumentation to alleviate the inflexibility of range tracking assets, increase their accuracy, and reduce operating costs. Several projects are planned to accomplish these tasks:

- Passive metric tracking system,
- Advanced landing systems using differential GPS,
- Automated range resource management system,
- Mobile launch head range system,
- Integrated launch head air-sea surveillance system, and
- Range dispersion monitoring system improvement.

Finding 3-25. Although large investments in range upgrades have been made over several decades, fundamental advancements in range functions and techniques remain elusive. NAI planning and resources appear to concentrate on improvements in existing systems. Real cost reductions will only come from a genuine paradigm shift in range operations.

Recommendation 3-19. NAI should continue to rethink the fundamental requirements and services provided by the range organizations. Experience gained in many years of military remotely piloted vehicle operations should be incorporated into the range concepts of operation.

Finding 3-26. The development of a secure advanced differential GPS landing aid will provide the most universal benefits applicable to all future RLVs.

Recommendation 3-20. NAI should concentrate appropriate resources to develop a secure advanced differential GPS landing aid.

The criticality of integrated range network architecture has been rated "medium" owing to the paradigm shift required to produce significant cost savings.

Intelligent Inspection

Systems to detect and isolate component or system failure have been in use for many years. Integrated vehicle health management (IVHM) systems not only perform those functions but also should identify part degradation and assist in vehicle maintenance. They could also enable a system to mitigate a failure by reconfiguring itself well enough to continue the mission or safely abort. An often used example is the detection of an impending engine failure during a multiengine assent. Failing engines may be throttled back (or completely shut down) prior to failure and other engines throttled up to compensate. The IVHM system could also notify the vehicle management computer (VMC) to alter the flight trajectory into a more benign environment. Development of such systems could help increase system safety and mission reliability. It could also greatly reduce operation costs by driving down scheduled maintenance and inspection hours. This technology will show tremendous benefits if and when it progresses to the point of predicting a premature component failure by examining real-time sensor data. Reaching this level of sophistication will require a significant investment. If the technology succeeds in this task, other subtle questions may surface. For example, will the decision authority authorize the next flight if the IVHM predicts the failure of a critical component on the second flight? Questions such as this should be fully examined in the course of this research. If successful, the payoff for developing IVHM systems could be substantial. Autonomous systems with embedded knowledge enhance operability while reducing highly skilled touch labor and enhance safety of hazardous operations and greatly enhance the overall responsiveness of the system.

Automated postflight inspection is another avenue toward significant cost savings. Specifically, reducing the time and manpower required for the inspection of large-acreage TPS and cryogenic assemblies could reduce the turnaround time and recurring costs of any RLV. These vehicles typically endure damage from both launch and orbital debris. TPS must be revalidated prior to each flight. Automatic detection and evaluation of impact damage via a scanning sensor would be of great benefit. The criticality of intelligent inspection has been rated "high" due to a significant potential reduction in turnaround time and postflight manpower requirements.

Security

Space systems require protection against unauthorized penetration of both facilities and electronic systems.

Finding 3-27. Although the security of space systems is vitally important to successful operations, very little security technology is unique to the space operations arena.

Recommendation 3-21. NAI should harvest and incorporate advances in security technology and techniques from sources outside the traditional aerospace community.

The criticality of security has been rated "low" since the technologies generated elsewhere can be imported as needed.

Summary

Launch operations is a key driver for total space lift costs. However, spending in some areas should be gated until a clear understanding of the final vehicle configuration emerges. Other technologies should be actively pursued to ensure they are ready when needed to realize the NAI program goals.

Mission Operations

The major NAI needs identified for mission operations are rapid mission planning (for the military), rapid response (for the military), on-orbit servicing and orbit transfer capability (both military and civilian), and affordable launch (both military and civilian). Affordable launch was discussed previously.

Rapid mission planning for military operations is required to respond effectively to threats. The need for situational awareness and for flexible systems that can adapt in near real time to changing mission requirements is increasing (Morrish, 2003). Space technologies for command, control, communications, computers, intelligence, surveillance, and reconnaissance (C4ISR) were identified as part of NAI but are beyond the scope of this access-to-space section. However, the ability to plan and launch missions effectively depends on global situational awareness and inputs from C4ISR. The ability to rapidly plan missions requires communications architectures matched to the situational awareness.

Efficient mission planning requires automated tools to trade off mission options. These tools must be capable of rapidly making complex trades between options for responding to threats. Tools are being used now by Operationally Responsive Spacelift (ORS) for long-range planning and launch vehicle requirements and evaluation. It is not clear if these tools are adaptable to real-time response planning—for example, which assets to launch in response to a given situation. NASA has extensive mission-planning tools for its missions, but these are either not designed for rapid response, or are extremely specialized to science missions. Mission planning tools must take into account surveillance and communications before the onset of hostilities, which space assets are capable of responsive launch and on-orbit checkout, the ability to launch on schedule (prehostilities), the ability to plan for persistent on-orbit presence, quick reconstitution of on-orbit assets, and quick launch to support tactical needs. These tools will need to be developed or upgraded from their current status.

Range tracking, that is the detection of the position and condition of vehicles during launch and landing, is currently a safety requirement. This function (see also "Launch Operations," above) is currently tied to fixed range systems for launch vehicles, but there is a desire to make the tracking of launch and landing independent of fixed locations to provide flexibility in launch and landing location and to facilitate rapid response (Morrish, 2003). The NASA tracking and data relay satellite system (TDRSS) and the Global Positioning System (GPS) were identified as the basis for such systems. DARPA's Responsive Access, Small Cargo, Affordable Launch (RASCAL) program includes an attempt to demonstrate this capability.

DARPA's RASCAL also proposes to demonstrate 24-hour mission turnaround and 1-hour scramble with aircraftlike sorties. RASCAL is to demonstrate a system that can launch microsatellites into any Earth-centered orbit without requiring fixed, ground-based ranges. A goal is to demonstrate mobile operations from any coastal airfield with a 5,000-ft runway and a system that can adapt in near real time to changing mission requirements.

Advances in orbit operations are required by both the Air Force and NASA to increase capability (e.g., for the Space Control mission) and, in turn, reduce costs. Both are focused on moving and servicing space assets. DARPA currently has a program called Orbital Express (it is not part of NAI), which is targeted for demonstrating the technical feasibility of robotic on-orbit servicing, robotic fueling, and orbital replacement unit (ORU) replacement by 2006. Operationally Responsive Spacelift (ORS) requires an orbital transfer capability to reposition/boost orbital assets and perform spacecraft servicing. The AFRL Space Vehicles directorate is working on a military space plane, which it defines as a space operation vehicle (booster), a space maneuver vehicle (a reusable transfer stage), a modular insertion stage (expendable for affordable space access), and a CAV (for Prompt Global Strike) (DDR&E, 2002).

DARPA's Orbital Express, plus NASA's planned work on an orbital maneuvering vehicle and its automated rendezvous and docking experience with the Russian Soyuz and Progress vehicles and others, make many of these technologies fairly mature. No funding has been identified for the XTV orbital transfer vehicle shown as part of NASA's Space Architecture, but the Space Transportation System and Russian technology base should make crewed XTV operations mature. Most existing DoD assets are probably not serviceable, so on-orbit servicing technology would only be applicable to certain future assets that show some economic or performance advantage to being designed for on-orbit servicing.

The technology requirements for on-orbit operations are focused on autonomy and associated technologies to make autonomous systems work. Specific technology needs are identified in Table 3-6.

Finding 3-28. Mission operations are hindered by the long-lead planning, antiquated range processes and requirements, and the arduous overhead that is imposed on space mission operations. These obstacles stem from a combination of factors—some are from 40 years of heritage and others from vehicle-unique constraints imposed by current vehicle designs.

Recommendation 3-22. Consider mission planning, quick response, and on-orbit operations as prime requirement drivers in the next series of vehicle designs. Specifically focus on an integrated program to develop the technologies needed for faster, more affordable operations.

Software

All aspects of the National Aerospace Initiative rely heavily on the use of software for successful applications. These include the three NAI major pillars: hypersonics, access to space, and space technology. One cannot imagine any aspect of this initiative that could be successfully completed without the use of modern-day computers and software. In fact, the software aspects often are one of the most costly items of the system life cycle. It is imperative that NAI reduce software costs and software errors if the initiative is to be successful.

The widespread, pervasive use of software throughout all programs related to the NAI is both a blessing and a burden to the program. The blessing is that there is much current work under way throughout the country, both in industry and academia, that can be used in the program. However, many of these programs must be focused specifically on NAI problems for NAI to be completely successful. This should not be difficult to make happen considering the high-level visibility of the NAI. Software evaluations are provided in Table 3-7.

Open Architectures

The criticality of open architectures was ranked "medium" because of the large body of work both in industry and in academia to solve these problems. Many of the presentations given to the committee listed these areas in the late 6.3 category, with TRLs of 6 or above.

Modularity

The probability of achieving the NAI goals of modularity is ranked "high." This is because much work on plug-and-play interfaces has already been done and is continuing. This work is in the late stages of development by industry and was covered in many of the presentations to the

TABLE 3-6 Status of Technologies for Mission Operations

NAI-Defined Technology Areas		Committee Evaluations			
		Constituent Technology Ratings			
Major Technology Area	Constituent Technology	Current Maturity Level (1-5)	Relative Impact on Major Tech Area (Negligible, Little, Moderate, Significant, Extreme)	Likelihood of Achieving in Time Frame (Current Funding Level) (Low, Medium, High)	Relative Criticality of Major Technology Area (Low, Medium, High)
Rapid mission planning—mission-unique constraints and analysis	Integrated mission planning tools	2	Moderate	Medium	High
	Mission unique constraints and analysis	1	Significant	Medium	
Rapid mission response	Range tracking	3	Moderate	High	Medium
	Launch and landing flexibility	2	Significant	Medium	
	Command and control	3	Moderate	Medium	
	Air traffic management	4	Little	High	
On-orbit operations	Autonomous GNC	3	Extreme	Medium	Medium
	Proximity operations	4	Significant	High	
	Grapple, soft dock	2	Moderate	High	
	Fuel/consumable transfer	2	Little	Medium	
	ORU transfer	1	Little	Medium	
	Orbit-assembly-compatible structures	4	Significant	High	
	Rapid sensor initialization	3	Extreme	Medium	

committee. This does not mean that some additional effort is not required to make these interfaces suitable for aerospace applications, but they are currently well developed generally speaking.

Engineering Design Software

Multiple engineering design software issues are very critical to the success of the NAI. The good news, however, is that there is a large body of work under way in all areas of engineering software. Several organizations presented their work to the committee. The areas presented covered combustion flow and analysis, simulation of aero control, stability and autopilot design, hypersonic inlet and nozzle design, thermal structural design, and propulsion system design. Most of these areas were being studied in the late 6.3 research category with moderate TRLs.

TABLE 3-7 Status of Technologies for Software

NAI-Defined Technology Areas		Committee Evaluations			
		Constituent Technology Ratings			
Major Technology Area	Constituent Technology	Current Maturity Level (1-5)	Relative Impact on Major Tech Area (Negligible, Little, Moderate, Significant, Extreme)	Likelihood of Achieving in Time Frame (Current Funding Level) (Low, Medium, High)	Relative Criticality of Major Technology Area (Low, Medium, High)
Open architectures	Secure wireless communications	3	N/A	High	Medium
Modularity	Plug-and-play interface	4	N/A	High	High
Engineering design software issues	Combustion flow analysis	3	Significant	Medium	Medium
	Simulation of aero control, stability, and autopilot design	4	Significant	Medium	
	Hypersonic inlet/ nozzle design	4	Significant	Medium	
	Thermal structural testing	4	Significant	Medium	
	Propulsion system design	2	Significant	Low	
Quick modification		3	N/A	Medium	Medium
Verification and validation	Parallel processing	2	Little	Low	High
	Prediction of vehicle performance	2	Significant	Low	
Autonomous flight control (transportable)	Vehicle to vehicle transportability	2	Moderate	Low	Medium
	Secure, wireless communications	4	Little	Medium	
	Automatic decision-making process	2	Significant	Low	
	Autonomous software for integration and test process	2	Little	Low	
Software engineering	Reliable, error-free codes	2	N/A	Low	High

Quick Modification

The committee was presented a reasonable amount of work addressing quick software modification of aerospace equipment and facilities. The work reported in this area had technology readiness levels of 5 and above and was therefore regarded as "medium" in criticality.

Verification and Validation

The relative criticality of software validation and verification (V&V) is ranked "high" because of the potentially catastrophic consequences of software errors and because of the extraordinary expense necessary to accomplish the task. Software verification is cumbersome and labor intensive. Research should not only concentrate on unique aerospace V&V issues but should also capitalize on the latest techniques being generated by the civilian software industries.

Autonomous Flight Control

Transportable, autonomous flight control has a number of critical areas to be addressed: vehicle-to-vehicle transportability, secure wireless communications, automatic decision processes, and software for autonomous integration and test processes. These areas are all in the early stages of applied research (6.2) with low TRLs. In particular, basic research should emphasize the automatic decision-making process.

Software Engineering

The software engineering area requires considerable work to develop reliable methods for generating and checking code. This is a very important area of research, with much work being done throughout the country, both in industry and academia. However, the state of development in this area is still judged to be low, and there are many opportunities for major breakthroughs if NAI gives it sufficient attention.

TECHNOLOGY EMPHASIS IN THE NEXT 5 TO 7 YEARS

Many areas of research require attention under the umbrella of the National Aerospace Initiative. This section delineates those that might be brought to bear on near-term applications (e.g., the military space plane or the next-generation launcher.) This is not to say that long-term technologies presented in the following sections should be discontinued—in fact, all technology presented in both groups must be pursued in parallel and given appropriate weighting.

Technologies having the greatest probability of enhancing the near-term goals of NAI are advanced materials for use in propulsion and thermal protection systems, electrical/hydraulic power generation and management, software transportability, integrated structures, and error-free software generation and verification. Furthermore, the development of computational analysis tools and methodologies should be emphasized—especially when coupled to test analysis and ground test facilities.

Propulsion research work should focus on all technologies contributing to engine reusability and reliability, such as the development of new engine materials that will support combustion of both hydrocarbon and hydrogen fuels. Additionally, advances in health management technologies would pay dividends in safety, extend engine life, and lower maintenance burdens.

Additional work in the materials area should focus on the needs of high-strength, LOx-compatible materials, lightweight durable thermal protection materials, and structural materials useful in new airframe designs. Durability should be the prime consideration in the development of these materials. The need for this work was emphasized in an earlier National Research Council report on hypersonics (NRC, 1998). Additional materials research must be focused on the need for reusable fuel tanks.

Both the DoD and NASA will benefit by leveraging work being developed by DOE for electrical power management and control. In this general area, advances in intelligent sensors and vehicle system thermal control should be sought by the NAI. Multiple programs being supported by the DoD and other agencies might be applicable.

Finally, the coupling of high-performance computing (numerical analysis) with emerging capabilities in ground test facilities (diagnostics, analysis tools, and methodologies) appears to show great promise in reducing the cost of vehicle development.

Finding 3-29. Milestones depicting the start of operational vehicle full-scale development have been established. The goal of NAI is to develop the underlying technologies to a level sufficient to support the decision milestones. The committee was not given a clear policy specifying the desired technology development thresholds at the milestones. Underlying technologies must be ready for full-scale engineering development at the appropriate decision points delineated in the NAI timetables—2008, 2015, 2018, and 2020 (Rogacki, 2003; Sponable, 2003). Significant flight test activity must also be completed prior to these decision points. Considering the short time between now and the end of Phase I in 2008, it is unlikely that the underlying technologies can be developed in time to meet the Phase I goals depicted in Figure 3-1.

Finding 3-30. The United States has failed to sustain many X-vehicle programs through completion of their flight test phase. Consequently, many years of technology advances have been prematurely halted and remain unproven in flight.

Recommendation 3-23. DoD and NASA should develop time-phased, reusable, rocket-based flight demonstration programs to move these and other near-term, unproven technologies through flight test; specify and disseminate the technology readiness levels and specific exit criteria necessary to support the operational decision points; ensure that research is directed toward obtaining the specified data and that the demonstrations—both flight and ground—are structured to obtain the required information and data; concentrate on technologies that contribute to reusability; and fund multiple copies of each design to mitigate potential loss of the entire program if a single vehicle is lost.

Finding 3-31. Advances in lightweight, high-strength materials—especially TPS, energy storage, and engine LOx-compatible materials—can make significant contributions to mass fraction and reusability.

Recommendation 3-24. DoD and NASA should strongly support basic and applied materials research programs. They should focus materials research in support of reliable, reusable, long-life, low-weight TPS, energy storage, and rocket engine systems.

Finding 3-32. Rocket-based RLVs are technically feasible in the near future and may provide lowest life-cycle costs for responsive, high launch rate requirements.

Finding 3-33. Achieving aircraftlike operations will require higher margin, more robust vehicles, and more efficient ground operations than current launch systems, as well as automated flight planning operations.

Recommendation 3-25. Adequate margin should be incorporated into all aspects of the next generation launch vehicle design. If necessary, payload capability should be sacrificed to achieve robust design goals.

Finding 3-34. Software is one of the most demanding and expensive aspects of any modern aerospace vehicle. Delays in software development ripple throughout the entire development process and significantly increase the overall cost of projects.

Recommendation 3-26. DoD and NASA should, through NAI, support a comprehensive research capability devoted to lowering the costs of aerospace software production. This research should concentrate on the safety-critical nature of aerospace software.

BUDGET SCENARIOS

The statement of task asks the committee to consider two budget scenarios for the development of NAI timelines: one that recognizes the current constrained Air Force budget, assuming no additional NAI funds are allocated, and one that meets the optimal NAI development timelines as developed by the committee. A rough order of magnitude estimate of the difference is also requested. The following paragraph addresses these requests only for the Air Force rocket-based access to space budgets.

Projected specific and related funding of AFRL space access efforts averages more than $100 million per year through FY 2009 without additional NAI-designated funds. This funding encompasses both upgrades to current expendable launch vehicles and technology for reusable systems. The NAI envisions a multiphase demonstration program with increasingly capable reusable vehicles available in 2009, 2015, and 2018. The development of these vehicles is not supported by current budgets. Additionally, while the first of these vehicles would essentially rely on current technology, the advances required for subsequent vehicles would require significant technology funding. The committee estimates that the first demonstration vehicle, the X-42, will cost $1 billion to $2 billion dollars and that subsequent phases will be increasingly expensive if they lead to an operational, high-rate, responsive launch system. In addition to the funding for the X-42, the committee believes it will be necessary to augment the space access technology budget by 100 to 200 percent to develop the requisite reusable systems capability for the Phase II and III systems.

LONG-TERM TECHNOLOGY AND PROGRAMS

Within the U.S. defense establishment, typical development periods from concept to initial operational capability (IOC) span 18 to 25 years (as they did for the C-17, F-22, FA-18 E/F, V-22, and others). Even the AIM-120 missile (AMRAAM), a relatively small system, took 12 years. This means that, with few exceptions, the concepts now emerging are close to what we can expect to be fielded—if and when the nation commits to improved spacelift capability and the improved effectiveness it provides to the LRS, ORS, and Space Control missions. Can these systems be developed faster and (consequently) cheaper? Yes. *Will* they be developed faster? Probably not without the impetus of a national crisis or some other cataclysmic precipitating event. Therefore, in the absence of unforeseen revolutionary technology advances, the fundamental difference between near- and far-term capabilities can be defined in terms of incremental advances in current research efforts. The results of this incremental process should not be underestimated. Slight changes in subsystem performance can drastically affect the vehicle configuration, operating characteristics, and consequently the life-cycle costs. A discussion of likely technology advancements follows.

Engine performance will undergo modest improvements in the medium term. A little more Isp and thrust-to-weight ratio can be squeezed out of chemical engines by making the mixture ratios leaner and incorporating lighter weight materials. With higher sortie rates, IVHM should mature to the point of making genuine contributions to safety and lower recurring costs. Higher sortie rates will also tip the economic scales toward automating current labor-intensive mission planning cycles. Aircraftlike ground and flight operations will become reality if vehicles are consciously allocated and constructed with substantial margin. Improvements in TPS materials will most likely result in robust, precipitation-tolerant flight characteristics. Advanced TPS materials may also enable durable, sharp leading edges and the accompanying improvements in maneuverability and operational flexibility they bring. Large-scale, high-Mach turbines might advance to the point of viability for launching low or possibly intermediate size payloads. However, their utility for launching large payloads will be severely challenged considering the extraordinary development and operating costs of multistaged, multifueled, multiengined, extremely heavy, horizontal take-off vehicles. Finally, the most significant advance will likely be seen in the way software is produced. It is likely that the extensive, worldwide resources devoted to software generation can be leveraged to form a development environment capable of producing timely, error-free code.

The military space plane has a good chance of becoming the next revolutionary/transformational/disruptive weapon system. Its development will only add to our already recognized dependency on space systems. Therefore, most space assets—including the MSP—will become valued targets in the eyes of our enemies. Concentrating on defensive maneuver, offensive engagement, and recovery from launch site attack will be necessary to counter the threat and capitalize fully on these emerging capabilities. The technology to produce an MSP currently exists or is very near term. Delaying the decision to field the MSP (while allowing other potentially useful technologies to mature) may or may not result in an overall lower cost system.

SUMMARY

The committee studied a variety of inputs and reference sources pertaining to the access-to-space pillar of the National Aerospace Initiative. Although NAI is chartered as a technology development effort, the pillar goals are expressed in terms of future operational system characteristics. This approach allows NAI to cast its net across a broad range of possible contributing technologies but necessitates the translation from system characteristics to technology objectives—a process open to some interpretation. Furthermore, without a clear understanding of the eventual system configuration, the specific system characteristics specified in the three-phase pillar could be difficult to justify and may result in expending resources on low-payoff technologies.

It is projected that the military space plane, as currently conceived, can perform the PGS, ORS, and Space Control missions in a very cost effective manner. The United States and several other countries currently possess the capability to develop the technology to produce this weapon system in the near term. However, over the last two decades, this country has lacked the conviction to demonstrate the underpinning technologies in flight. A flight demonstration is the next logical step to develop this transformational capability. It is yet to be seen if the emerging Air Force interest in these missions will produce tangible results.

Summarized in this chapter is the committee's analysis and opinion of the NAI plan and the critical technologies that are required to enable the near- and long-term goals of the NAI. In general, the committee finds the NAI space access goals to be operationally relevant, technically feasible, and underfunded to meet the proposed schedule. This chapter has been prepared so that the NAI participants can compare these findings and recommendations with established plans and programs. The recommendations here are therefore intended to serve as input for evaluation.

REFERENCES

Published

Antonio, M. 2003. Launch services: Too many rockets, too few payloads. Aviation Week and Space Technology. Available at http://www.aviationnow.com/avnow/news/channel_awst_story.jsp?id=news/sb03_12.xml. February 24.

Boeing. 2001. Rocketdyne receives $65 million from NASA for next-generation space propulsion development. News release found at http://www.boeing.com/news/release/2001/q2/news release 010522s.html. Accessed on January 6, 2004.

DoD (Department of Defense). 2000. Department of Defense Space Technology Guide, FY 2000-01. Available at http://www.defenselink.mil/nii/org/c3is/spacesys/STGMainbody.pdf. Accessed on January 6, 2004.

NRC (National Research Council). 1998. Review and Evaluation of the Air Force Hypersonic Technology Program. Washington, D.C.: National Academy Press.

Singer, J. 2003. New test facility to save time, money on RASCAL program. SpaceNews. November 3, p. 9.

Williams, K. 2000. Small Companies to Study Potential Use of Emerging Launch Systems for Alternative Access to Space Station. Available at http://www.spaceref.com/news/viewpr.html?pid=2467. Accessed on January 7, 2004.

Unpublished

Brockmeyer, J.W. 2003. Integrated High Payoff Rocket Propulsion Technologies (IHPRPT). Briefing by Jerry Brockmeyer to a fact-finding subgroup of the Committee on the National Aerospace Initiative. September 29.

Candler, G.V. 2003. Synthesizing Numerical Analysis Data to Design and Develop Hypersonic Engines and Airframes. Briefing by Graham Chandler, University of Minnesota, to the Committee on the National Aerospace Initiative. October 7.

DDR&E. 2002. National Aerospace Initiative. Access to Space Draft S&T Plan. Rosslyn, Va.: DDR&E. September.

DDR&E NAI. 2003. Space Access Taxonomy. Handout given to committee to understand GOTChA process for space access. Rosslyn, Va.: DDR&E, October 22.

DeGeorge, D. 2003. Intersection of Air Force Propulsion S&T and the National Aerospace Initiative. Briefing by Drew DeGeorge, AFRL, to the Committee on the National Aerospace Initiative. September 3.

Holden, M.S. 2003. Requirements and Facilities for Ground Test of Full-Scale Scramjet Engines at Duplicated Flight Conditions from Mach 8 to 15. Briefing by Michael Holden, CUBRC, to the Committee on the National Aerospace Initiative. October 7.

Kastenholz, C. 2003. National Aerospace Initiative (NAI) Payloads. Briefing by Lt. Col. Chuck Kastenholz, AFRL, to the Committee on the National Aerospace Initiative. September 3.

Koozin, W. 2003. ORS Operations Concept. Briefing by Walt Koozin, AFSPC, to the Committee on the National Aerospace Initiative. October 7.

Lyles, G. 2003. NGLT Briefing to the Air Force Science and Technology Board. Briefing by Garry Lyles, NASA Marshall Space Flight Center, to the Committee on the National Aerospace Initiative. October 7.

McClinton, C. 2003. NASA Hypersonics. Briefing by Charles McClinton, NASA Langley, to the Committee on the National Aerospace Initiative. September 4.

Morrish, A. 2003. DARPA Programs for the National Aerospace Initiative Objective. Briefing by Art Morrish, DARPA, to the Committee on the National Aerospace Initiative. August 5.

Rogacki, J. 2003. The National Aerospace Initiative—A Synergistic Partnership Between NASA and DoD. Briefing by John "Row" Rogacki, NASA, to the Committee on the National Aerospace Initiative. August 5.

Sega, 2003. National Aerospace Initiative—NRC Briefing. Briefing by Ronald M. Sega, DDR&E, to the Committee on the National Aerospace Initiative. August 6.

Sponable, J. 2003. National Aerospace Initiative Overview. Briefing by Jess Sponable, AFRL, to the Committee on the National Aerospace Initiative. September 3.

4

Enabling a More Robust Aerospace Workforce

INTRODUCTION

One of the committee's tasks was to suggest initiatives that ensure a more robust aerospace science and engineering workforce will be available to develop hypersonic propulsion, access to space, and space technology to meet civilian and warfighter needs over the next 20 years.

The Hart-Rudman Commission illustrated the relation between U.S. national security and the need for a national effort in aerospace science and technology (S&T) that is supported by a well-educated aerospace workforce. The report stated in blunt terms,

> [t]he harsh fact is that the U.S. need for the highest quality human capital in science, mathematics, and engineering is not being met. . . . This [situation] is not merely of national pride or international image. It is an issue of the utmost importance to national security. In a knowledge-based future, only an America that remains at the cutting edge of S&T will sustain its current world leadership. . . . Complacency with our current achievement of national wealth and international power will put all of this at risk. (USCNS, 2001)

America's aerospace leadership depends on the products of aerospace, which in turn depend on a continuous supply of scientists and engineers engaged in work funded from the 6.1 to 6.7 level. At present, we do not have a national aerospace consensus to help guide policies and programs. This contributes to unfocused federal and industrial investments spread over a variety of long-range research programs and their associated aging infrastructure. The DoD has based its continuing military dominance on technological superiority rather than on maintaining large numbers of personnel in the uniformed services. This is a strategy that has worked well to now. However, our current military dominance was derived from long-term investments in S&T made in the 1950s through the 1970s by the DoD and other federal agencies, including the National Aeronautics and Space Administration (NASA). Defense capabilities that are now entering service are based on the S&T investments of the 1980s and 1990s. There is concern that

> [t]he focus of the current DoD S&T program is primarily on incremental improvements in current capabilities . . . and does not place sufficient emphasis on innovative technology initiatives leading to entirely new military capabilities. (DSB, 2000)

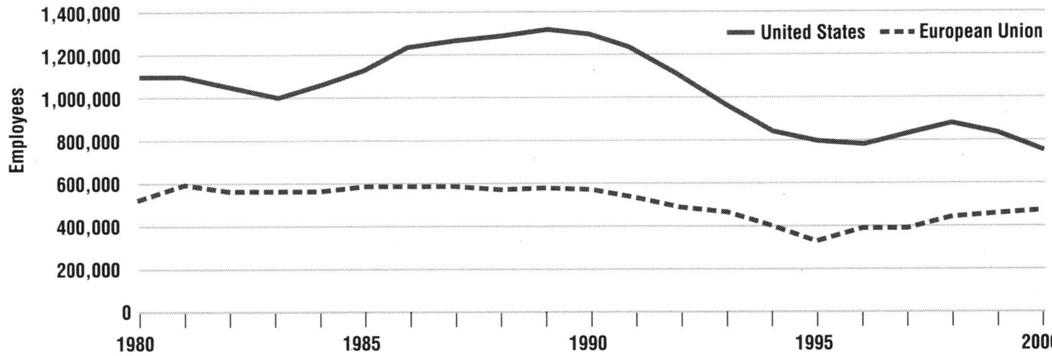

FIGURE 4-1 U.S. and EU aerospace employment, 1980-2000. SOURCE: CFUSAI, 2002.

In the long term, this situation is untenable. As other nations build their own aerospace workforces, the current balance will inevitably shift, eroding or even ending our military advantage. This is already happening in the commercial aerospace sector.

> Our policymakers need to acknowledge that the nation's apathy toward developing a scientifically and technologically trained workforce is the equivalent of intellectual and industrial disarmament and is a direct threat to our nation's capability to continue as a world leader. (CFUSAI, 2002)

The following are some of the trends characterizing the current state of the aerospace workforce of the United States:

- A decline of 60 percent from the 1964 peak in the overall federal investment in R&D as a percentage of GDP, in particular for the physical sciences and engineering, even though the DoD depends heavily on the application of these fields to develop new warfighting and peacekeeping capabilities.
- International competitors, including the European Union, are gaining (Figure 4-1). The U.S. share of the world's aerospace markets as measured by annual revenue fell from over 70 percent to below 50 percent in 2000 (CFUSAI, 2002).
- Because of industry layoffs that were due to the end of the Cold War and subsequent mergers, the nation has lost over 600,000 scientific and technical aerospace jobs since 1989, with 100,000 lost since the attacks of September 11, 2001 (Figure 4-2). Many of those laid off move on to other industries, which include nontechnical industries like banking and entertainment (CFUSAI, 2002; NASA, 2003a).
- The average age of the aerospace worker in industry is 44. The average age is 51 at NASA and 53 in the DoD.[1] Over 26 percent of the aerospace workforce will be eligible for retirement in 2008 (CFUSAI, 2002).
- The proportion of aerospace workers 30 years old or younger dropped from 18 percent in 1987 to 6.4 percent in 1999 (AIA, 2001). In part this is due to the cyclical nature of major aerospace programs, which caused a general decline in the number of American-born youth pursuing an education in aerospace engineering at the undergraduate level (Figure 4-3) and the graduate level (Figure 4-4) (McLaughlin, 2002; Grossman, 2002; CFUSAI, 2002; NASA, 2003a). Another reason

[1] Statement of Thomas Kochan, Codirector, Labor Aerospace Research Agenda, Massachusetts Institute of Technology, before the Commission on the Future of the United States Aerospace Industry, May 14, 2002.

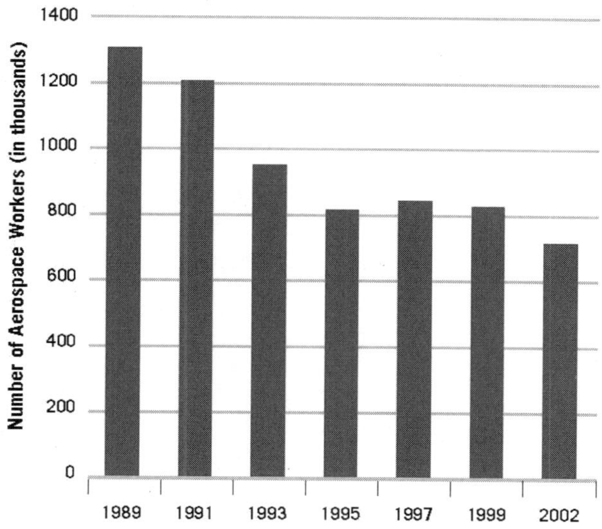

FIGURE 4-2 Total employment in the aerospace industry, 1989-2002. SOURCE: CFUSAI, 2002.

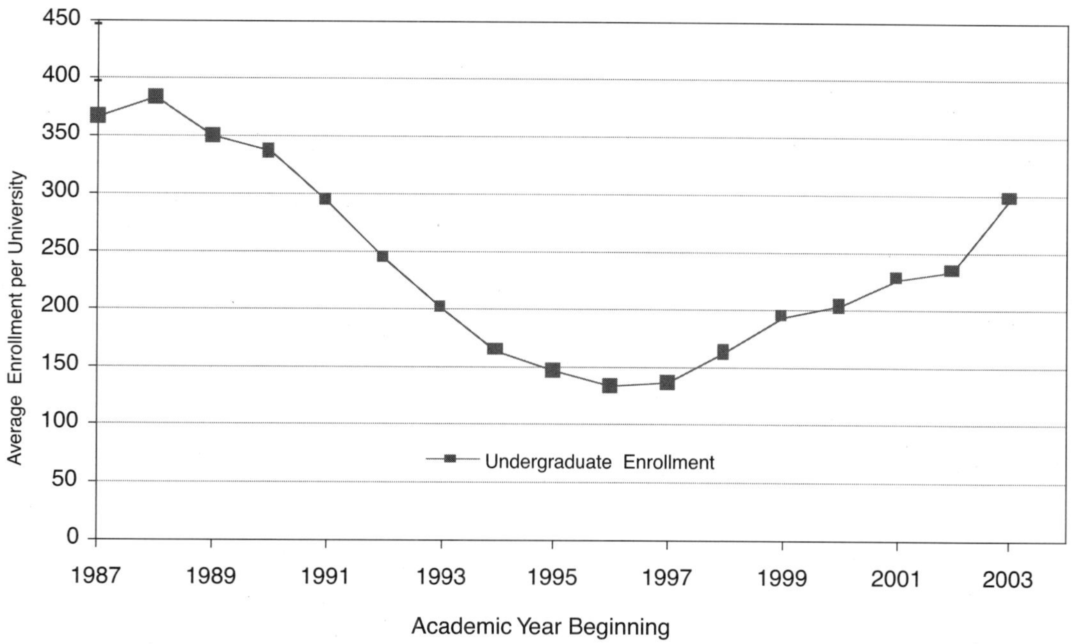

FIGURE 4-3 Average aerospace undergraduate enrollment per university (37 universities), 1987-2003. SOURCE: McGlaughlin, 2002, as updated via e-mail.

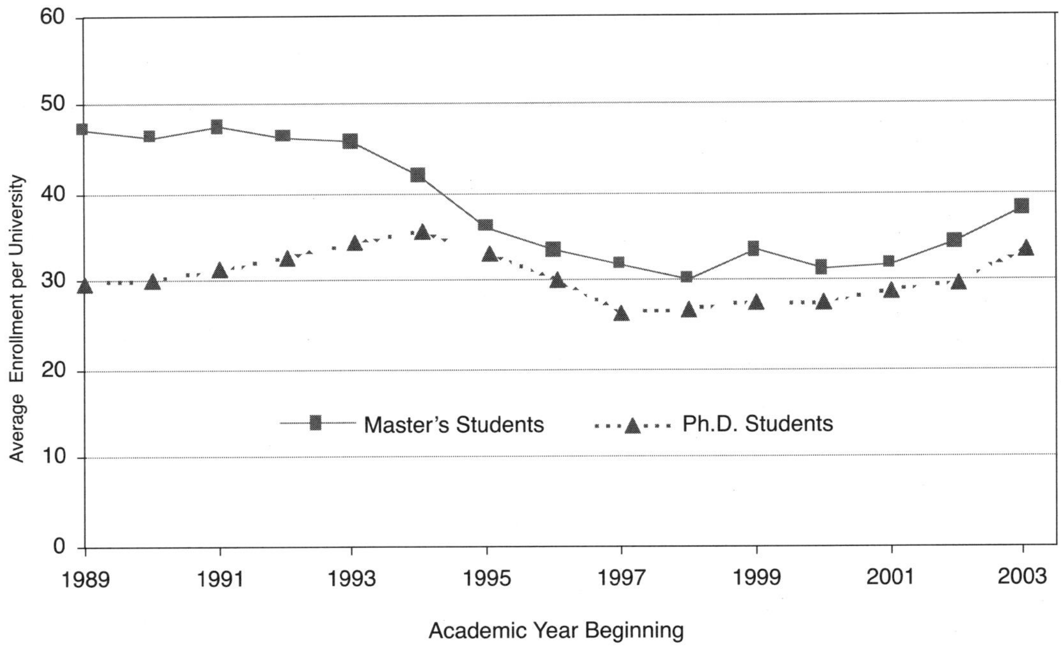

FIGURE 4-4 Average aerospace graduate enrollment per university (37 universities), 1989-2003. SOURCE: McGlaughlin, 2002, as updated via e-mail.

may be the failure of mathematics and science education in the K-12 system (Figure 4-5), which may be a more pertinent reason for declining engineering enrollments and not unique to the aerospace discipline (Figures 4-6 and 4-7).

The U.S. workforce has always benefited from international S&E talent (Figure 4-8). However global competition for this talent is intensifying. The number of doctorates earned by U.S. citizens in engineering has decreased from 2,514 in 1970 to 2,206 in 2000. At the same time, the number of doctorates earned by foreign students on temporary visas grew from 471 in 1970 to 2,444 in 2000. An increasing and significant percentage of these highly educated scientists and engineers enter the

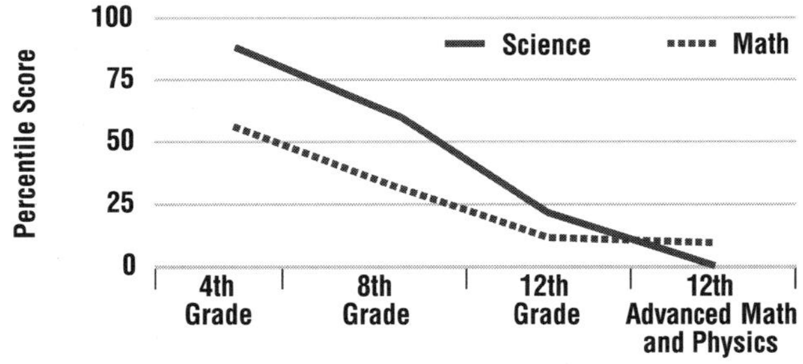

FIGURE 4-5 Science and math performance of U.S. students relative to foreign students, 2001. SOURCE: CFUSAI, 2002.

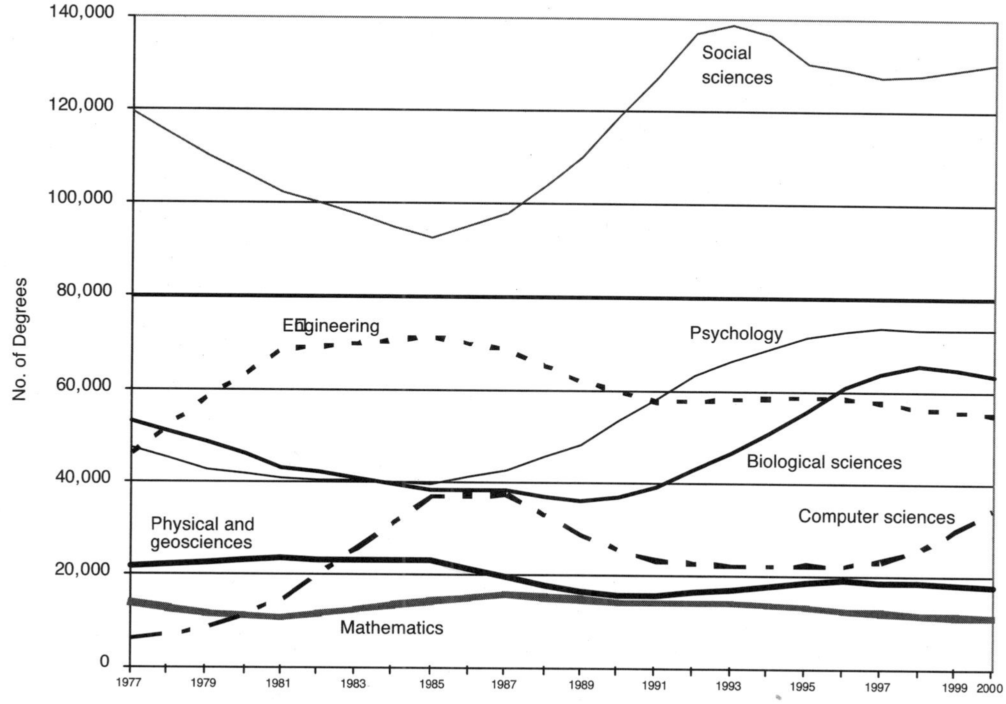

FIGURE 4-6 B.S. degrees earned in selected science and engineering fields by U.S. citizens and permanent residents, 1977-2000. SOURCE: NSB, 2003.

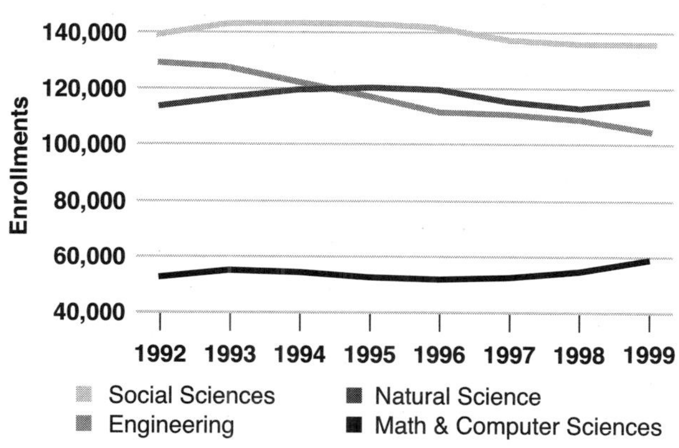

FIGURE 4-7 Graduate enrollment in science and engineering fields, 1992-1999. SOURCE: CFUSAI, 2002.

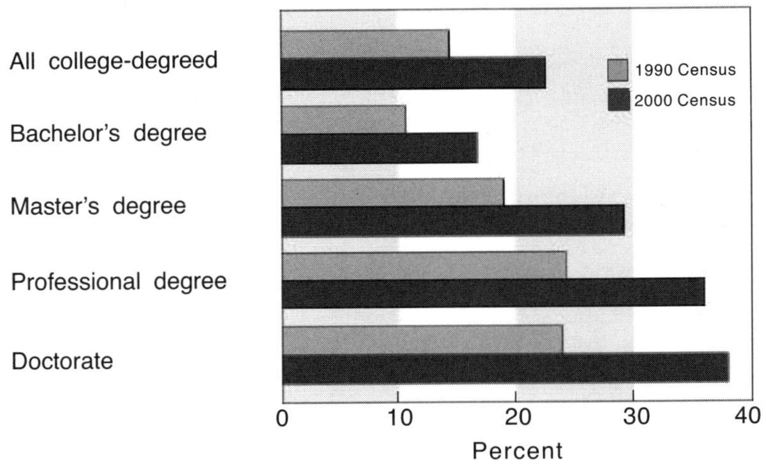

FIGURE 4-8 Share of foreign-born scientists and engineers in U.S. S&E occupations by degree level, 1990 and 2000. SOURCE: NSB, 2003.

aerospace workforce of their home countries; we are educating our own competition. Many of those who do stay in the U.S. cannot obtain security clearances required for NAI-relevant aerospace positions. As a result, a considerable number of scientists and engineers educated in our universities do not or cannot contribute to this nation's aerospace efforts.

NASA and DoD representatives presented to the committee their approaches to aerospace workforce development. Though efforts are being made to address workforce concerns, the committee believes these are insufficient. The committee believes what is really needed to address the majority of our workforce challenges is a clearly stated national aerospace consensus that focuses long-term federal and industrial investments on projects of strategic interest. This would include establishing a modern version of the National Defense Education Act of the Sputnik era that could be coordinated through an interagency task force. The benefits would cascade throughout the federal, industrial, and educational systems. A national aerospace consensus with consistent and long-term funding would mitigate the cyclical nature of hirings and layoffs and foster the creation of stable, career-oriented jobs of strategic importance. This, in turn, with the aid of an interagency task force, would attract high-quality undergraduate and graduate students who were inspired as K-12 students. Evidence of the effect of national efforts and the corresponding increase in funding on undergraduate and graduate education can be seen in Figures 4-3 and 4-4, respectively. It is no coincidence that increases in enrollment in aerospace programs coincided with increases in DoD spending and industry demand in the late 1980s and those beginning in the late 1990s.

The next two sections in this chapter discuss the two findings of the committee on the aerospace workforce and the respective recommendations. Finding 4-1 explains the DoD and NASA attempt to address the problem of the rapidly aging aerospace workforce and the resulting loss of technical knowledge by implementing a human capital management system. The discussion that follows Finding 4-2 explains the attempts that are being made to address education of the future aerospace workforce through basic research and collaboration between universities, industry, and government laboratories.

REINVIGORATING THE WORKFORCE

Finding 4-1. As mentioned in the introduction, the average age of the aerospace worker (industry, NASA, and DoD) is 49, and approximately 26 percent of this same aerospace workforce will be eligible for retirement in 2008. Upon their retirement, these aerospace workers will take with them specialized knowledge accumulated over decades. Loss of technical knowledge is most acute in the air-breathing hypersonics community. As a member of the NAI, NASA is developing a stopgap measure that will implement an integrated agency-wide approach to human capital management. This approach is designed to attract and maintain a workforce that is representative of the nation's diversity and includes the competencies that NASA needs to deliver sustained levels of high performance that its challenging mission requires (NASA, 2003a). The DoD works to identify mission-critical occupations and competencies needed in the current and future workforce and to develop strategies to identify, recruit, and retain a high-performing workforce (OPM, 2004).

Discussion 4-1. NASA's personnel reduction impacted its younger staff more severely. During the start of the NASA downsizing in 1993, the number of S&T workers under the age of 30 was nearly double the number of those over 60. At the conclusion of the downsizing, the number of personnel under the age of 30 was one-third the number of those over 60. Individuals with critical skills were lost (NASA, 2003a). In addition, almost 15 percent of NASA's workforce is eligible for retirement now, and 25 percent is eligible within 5 years. Some core competencies can easily be lost with the retirement of only a few individuals. (This is true within both industry and the DoD.) To address this and other workforce trends, NASA formed the agency-wide Strategic Human Capital Plan and the Strategic Human Capital Implementation Plan (NASA, 2003a; NASA, 2003b). These plans are meant to identify those civil servants needed for mission-critical competencies that NASA must have in-house and to implement initiatives that capture corporate knowledge in an effective and systematic way from those who are retiring from the organization.

NASA's approach does not address the long-term problem the nation faces in the 21st century. To maintain aerospace leadership, the United States must continue to have a highly skilled, stable, secure, and growing aerospace workforce. Many young Americans, however, are rejecting careers in aerospace in favor of more lucrative professions. Undergraduate and graduate students must find a rewarding future in aerospace.

Students interested in research see the prospect of low-paid apprenticeships, training that requires a decade or more, modest stipends for graduate and postdoctoral students, who are often in their mid-30s, and limited job opportunities. More importantly, prospects for autonomous research positions in academia and related research-intensive employment opportunities that most would-be scientists aspire to at the end of their long road are uncertain and increasingly slim (Zumeta and Raveling, 2002). Anecdotal evidence points to faculty having particular difficulty retaining students in air-breathing hypersonics. If one mentions the exciting advances that have come out of the joint DoD-NASA National Aerospace Plane (NASP) program—NASA's X-33, X-34, and X-37, and the DARPA affordable rapid response missile demonstrator (ARRMD) (DDR&E, 2003)—it will be pointed out that all of the projects were canceled and not one of the X vehicles has flown.

Recommendation 4-1a. The aging workforce problem is one shared by all of aerospace and cannot be addressed by NAI alone, but NAI can contribute to the solution. An aerospace workforce pillar should be established in the NAI, along with the air-breathing hypersonics, space access, and space technology pillars. This workforce pillar can assist in a long-term national effort to coordinate all aerospace interests with consistent and long-term funding, as recommended in the introduction to this chapter. In the long term, an aerospace workforce pillar would allow much greater flexibility, coordination, and collaboration between industry, government, and universities than are presently found in the URETIs. As an added benefit, the work could be managed from one office and coordinated with the efforts of the interagency task force, also proposed in the introduction to this chapter.

Recommendation 4-1b. The NAI pillar embodied in Recommendation 1a should exercise creativity and innovation to form extensive cooperation between academia, industry, and federal agencies to preserve the aging workforce's technical knowledge. For example, the science, mathematics, and engineering (SME) educational programs that are funded by the DDR&E (EDUGATE, 2004) can be designed so that older aerospace workers can exchange knowledge and experience with local universities. Also, NASA and the DoD could augment their civil service and military personnel with an equal number of university, nonprofit, and industry personnel (NRC, 2001). This term-limited mixing in of top industry and academic talent would provide DoD employees with broader exposure to new ideas and help preserve DoD corporate memory.

Recommendation 4-1c. In the short term, NASA should be allowed to implement some of the human capital legislative options to address the aging workforce problem, as detailed in Appendix B of the Strategic Human Capital Implementation Plan (NASA, 2003b).

Recommendation 4-1d. A detailed study on the size of the S&E workforce required to meet future needs of the NAI should be commissioned after the initiative's long-term goals and funding have been solidified.

UNIVERSITY RESEARCH, ENGINEERING, AND TECHNOLOGY INSTITUTES

Finding 4-2. The NAI recognizes that a balanced investment portfolio, from basic research to technology demonstration, is essential to fulfilling its goals and to countering the previously mentioned trends in aerospace workforce education (DDR&E, 2003). As a response, two URETIs have been jointly funded by the DoD and NASA (DDR&E, 2003).

Discussion 4-2. The two URETIs are led by the University of Maryland,[2] investigating third-generation reusable launch vehicles, and the Georgia Institute of Technology, investigating aerospace propulsion and power for NASA's long-term aerospace enterprise goals (Zinn, 2003). From information obtained by the committee, the URETIs include undergraduate and graduate student collaboration with university and industry partners and access to national facilities, including the NASA Glenn Research Center, the NASA Langley Research Center, the NASA Marshall Space Flight Center, the Air Force Research Laboratory, the Naval Air Warfare Center, and the Arnold Engineering Development Center. The URETIs address student research symposia, scholarships and fellowships, K-12 outreach programs, outreach to women and underrepresented minorities, and exchange of faculty, industry, and NASA researchers. Most importantly, the basic research aligns fully with the NAI goals of developing reusable vehicles for space access.

In general, URETIs are attractive from the perspective of NASA and the DoD as they are relatively easy to manage and can make considerable progress in getting new research off the ground by attempting to provide the critical mass of students, the needed cross-fertilization between university, industry, and government researchers, and dependable multiyear funding. However, there are drawbacks to applying the URETI model to the NAI components, especially in the development of air-breathing hypersonic launch vehicles.

---

[2] Personal communication from Mark Lewis, University of Maryland, to Kenneth Harwell, DDR&E, October 16, 2003.

• The NAI-affiliated URETIs operate on a winner-take-all basis. By forming these two URETI centers, NASA and the DoD have provided the faculty with the largest share of support in the area of air-breathing hypersonics. These two URETI centers cannot, however, provide the entire solution to the basic science problems inherent in hypersonics. The aerospace community has a paucity of experts in the complex problems associated with air-breathing hypersonics and should not exclude those not affiliated with member universities.

• Funding for NAI URETIs, at $3 million per year, is not sufficient to form a critical mass of faculty and students to tackle the major research problems in reusable launch vehicles.

• The NAI URETIs may be working on an overly difficult problem. The development of air-breathing launch vehicles is the most difficult of the NAI components. URETIs are usually funded for a 5- to 7-year period with the option of renewal. This is too short a time period to demonstrate significant progress in the hypersonics problem.

• Both of the two NAI-selected URETIs appear to have K-12 outreach programs in their charters, but no matter how great the effort, they will have a limited, local influence. The NAI needs a much broader range of academic institutions from which to draw its workforce.

Recommendation 4-2a. The NAI should clearly articulate a long-term commitment to reusable launch vehicle research. Adequate and consistent federal R&D funding to this problem is far more important than management convenience. Funding concerns can be mitigated by collaboration with other science and engineering fields. Most of the elements that must be addressed in the NAI require a multidisciplinary approach, using tools from computational and experimental fluid dynamics, uncertainty analysis, MDO, and materials science. Advances made with any of these tools will benefit all of S&T.

Recommendation 4-2b. NAI should encourage lifelong learning and individualized instruction, especially in government laboratories, to attract and retain aerospace workers (NRC, 2003). This can most easily be done through existing education programs that are already funded by DDR&E (EDUGATE, 2004). These programs can be expanded beyond their respective directorates and centers to other NAI-relevant laboratories.

Recommendation 4-2c. In the near term, the NASA-funded URETI led by the University of Florida on third-generation reusable launch vehicles should also be affiliated with the NAI (UF, 2003).

Recommendation 4-2d. In the near term, the URETI concept should be modified and expanded to form a critical mass of faculty and students in the air-breathing hypersonics component of NAI without excluding acknowledged experts in the field. For example, university centers of excellence (ARL, 2004) could be formed around experimental and computational facilities or core expertise through competitive grants to individual investigators.

Recommendation 4-2e. In the near term, the NAI should establish specific graduate fellowship and undergraduate scholarship programs with outreach to women and minority groups that are underrepresented in aerospace.

Recommendation 4-2f. In the near term, funds should be added to existing high school and undergraduate internship and faculty fellowship programs at NASA and DoD laboratories that work on NAI-relevant research (EDUGATE, 2004; ASEE, 2004). This would give high school students and aerospace undergraduates across the nation exciting hands-on experience, often lacking in many curricula. These same interns could also be charged with becoming "engineering ambassadors" in secondary school endeavors (UA, 2003, 2004; KSU, 1998). Faculty members would carry their experience back to the classroom.

REFERENCES

Published

AIA (Aerospace Industries Association). 2001. Attracting and Retaining Qualified Personnel in the Aerospace Industry.

ARL (Army Research Laboratory). 2004. Partnerships. Program available at http://www.arl.army.mil/main/Partnerships/default.cfm?Action=partnerships&header=YES. Accessed on March 24, 2004.

ASEE (American Society for Engineering Education). 2004. Program available at http://www.asee.org/fellowship/default.cfm. Last accessed on March 24, 2004.

CFUSAI (Commission on the Future of the United States Aerospace Industry). 2002. Final Report of the Commission on the Future of the United States Aerospace Industry. Available at http://www.aerospacecommission.gov/AeroCommissionFinalReport.pdf. Accessed on December 23, 2003.

DDR&E. 2003. The National Aerospace Initiative Office, Congressional Report on the National Aerospace Initiative: In Response to the Senate Armed Services Committee Report Accompanying the National Defense Authorization Act for Fiscal Year 2004 (S.R. 108-46), September.

DSB (Defense Science Board). 2000. Letter Report on the Adequacy of the DoD Science & Technology (S&T) Program. Available at http://www.acq.osd.mil/dsb/letter.pdf. Accessed on March 24, 2004.

EDUGATE. 2004. Welcome to Edugate webpage. Program available at http://web.lmi.org/edugate/information.html. Accessed on March 24, 2004.

Grossman, B. 2002. Aerospace Engineering Enrollment and Employment Trends. Available at http://www.aerospacecommission.gov/051402testimony/grossman.pdf. Accessed on March 24, 2004.

KSU (Kansas State University). 1998. New engineering ambassadors ready to visit local high schools. eCollegian 102(108). Available at http://www.kstatecollegian.com/issues/v102/sp/n108/news/cam-ambass-williams.html. Accessed on March 24, 2004.

NASA (National Aeronautics and Space Administration). 2003a. NASA Strategic Human Capital Plan. Available at http://nasapeople.nasa.gov/hcm/Agency_SHCP-Final1.pdf. Accessed on March 24, 2004.

NASA. 2003b. Strategic Human Capital Implementation Plan. Available at http://nasapeople.nasa.gov/hcm/Agency_SHCP-Final2.pdf. Accessed on March 24, 2004.

NRC (National Research Council). 2001. Review of the Future of the U.S. Aerospace Infrastructure and Aerospace Disciplines to Meet the Needs of the Air Force and the Department of Defense. Washington, D.C.: National Academy Press.

NRC. 2003. Envisioning a 21st Century Science and Engineering Workforce for the United States: Report to the Government-University-Industry Research Roundtable. Washington, D.C.: The National Academies Press.

NSB (National Science Board), Committee on Education and Human Resources. 2003. The Science and Engineering Workforce: Realizing America's Potential. Available at http://www.nsf.gov/nsb/documents/2003/nsb0369/start.htm. Accessed on March 24, 2004.

OPM (Office of Personnel Management). 2004. Strategic Management of Human Capital. Program available at http://www.opm.gov/Strategic_Management_of_Human_Capital/index.asp. Accessed on March 24, 2004.

UA (University of Arizona), College of Engineering and Mines. 2003. Engineering ambassadors. Arizona Engineer 26(2): 3.

UA, College of Engineering and Mines. 2004. Community Outreach: Engineering Ambassadors Webpage. Available at http://www.engr.arizona.edu/2_CO/engamb.htm. Accessed on March 24, 2004.

UF (University of Florida). 2003. Institute for Future Space Transport. Program available at http://www.mae.ufl.edu/ureti. Accessed on March 24, 2004.

USCNS (United States Commission on National Security/21st Century). 2001. Roadmap for National Security: Imperative for Change. The Phase III Report. Available at http://www.nssg.gov/PhaseIIIFR.pdf. Accessed on March 24, 2004.

Zumeta, W., and J.S. Raveling. 2002. Attracting the best and brightest. Issues in Science and Technology. Available at http://www.evansuw.org/FAC/Zumeta/pdf/attracting_the_best.pdf. Accessed on March 24, 2004.

Unpublished

McLaughlin, D.K. 2002. ADCA/AIAA Aerospace Engineering Enrollment Survey, Fall 2001. Presented by Dennis K. McLaughlin at the AIAA Aerospace Sciences Meeting, Reno, Nevada. January.

Zinn, B.T. 2003. NASA/DoD UAPT Program Overview. Briefing by Ben Zinn, Georgia Institute of Technology, at NASA Glenn Research Center. Available at http://www.asdl.gatech.edu/teams/ureti/Combust_Pres/GT_URETI_Program_GRC_Visit.pdf. Accessed on March 24, 2004.

Appendixes

A

Biographical Sketches of Committee Members

Edsel D. Dunford (NAE), *Chair*, is the past president and chief operating officer of TRW, Inc. Over a 35-year career with Boeing Airplane Company, Ford Aerospace, and TRW, Inc., he focused on spaceborne electronics and the management of technology programs. Mr. Dunford holds a B.S. in electrical engineering from the University of Washington and an M.S. in engineering from the University of California at Los Angeles. Mr. Dunford was recently the coexecutive producer of a 2-hour historical documentary, "The Cold War and Beyond," which aired in fall 2003 and was a finalist at the Hollywood Film Festival. His experience is in military acquisition and procurement, aerospace engineering, space science, strategic planning, and systems engineering.

Donald J. Kutyna (U.S. Air Force, retired), *Vice-Chair*, is currently the vice president of Space Technology at Loral Space and Communications. Prior to this position, he served as vice-president of Advanced Space Systems at the Loral and Lockheed Martin Corporations. General Kutyna retired from 35 years of service in the Air Force. In his final assignment he served simultaneously as commander in chief (CINC) of the North American Aerospace Defense Command, CINC of the United States Space Command, and commander of the Air Force Space Command. In these positions, he was responsible for the acquisition, operation, and maintenance of all DoD space systems and supporting space assets. He also chaired the Technical Investigation Panel of the Space Shuttle Challenger Presidential Commission. General Kutyna holds a B.S. degree from the United States Military Academy and an M.S. in aeronautics and astronautics from the Massachusetts Institute of Technology. He has experience in military operations, acquisition program management, aerospace engineering, space science, propulsion engineering, strategic planning, systems engineering, orbital mechanics, aerodynamics, military and commercial space satellites, space mission success, space operations and missile warning, and space launch and control.

Kevin G. Bowcutt is a Boeing Senior Technical Fellow and chief scientist of hypersonic design and applications for the Boeing Company. Most recently, Dr. Bowcutt led the design team that created the Flexible Aerospace System Solution for Transformation (FASST) two-stage-to-orbit, air-breathing, reusable launch vehicle concept that is now part of the NASA Next-Generation

Launch Technology program. He previously worked for the National Aerospace Plane (NASP) program, led a project testing scramjet engines at Lawrence Livermore National Laboratory, and led the conceptual design efforts for the DARPA/Boeing affordable rapid-response missile demonstrator (ARRMD) vehicle. Dr. Bowcutt holds B.S., M.S., and Ph.D. degrees in aerospace engineering from the University of Maryland, College Park. He has experience in aerospace engineering, physics, thermodynamics, propulsion engineering, systems engineering, strategic planning, aerodynamics, program management, propulsion integration, and integrated vehicle design and optimization.

Kenneth E. Eickmann (U.S. Air Force, retired) spent the last 5 years as director of the Construction Industry Institute (CII) at the University of Texas. Prior to joining the CII, he completed a 31-year career in the U.S. Air Force, in which his last assignment was commander of the Aeronautical Systems Center at Wright-Patterson Air Force Base, where he led the nation's largest center of excellence for research, development, and acquisition of aircraft, aeronautical equipment, and munitions. Lieutenant General Eickmann holds a B.S. in mechanical engineering from the University of Texas, Austin, and an M.S. in systems engineering from the Air Force Institute of Technology. He is also a senior lecturer in the College of Engineering at the University of Texas at Austin and a registered professional engineer. His expertise is in military acquisition and procurement, aerospace engineering, materials science and engineering, government technical program management, propulsion engineering, strategic planning, industrial engineering, and systems engineering.

Wesley L. Harris (NAE) is the Charles Stark Draper Professor and head of the Department of Aeronautics and Astronautics at the Massachusetts Institute of Technology. His research focuses on theoretical and experimental unsteady aerodynamics and aeroacoustics; computational fluid dynamics; and the impact of government policy on procurement of high-technology systems. Prior to this position he served as the associate administrator for aeronautics at NASA. He has also served as the vice president and chief administrative officer of the University of Tennessee Space Institute. Dr. Harris earned a B.S. in aerospace engineering from the University of Virginia and M.S. and Ph.D. degrees in aerospace and mechanical sciences from Princeton University. His experience is in military acquisition and procurement, aerospace engineering, government technical program management, thermodynamics, propulsion engineering, strategic planning, systems engineering, and aerodynamics.

Hans G. Hornung (NAE) is the Clarence L. Johnson Professor of Aeronautics at the California Institute of Technology. His research focuses on turbulence and flow in hypersonic environments. Formerly he was the director of the DFVLR Institute for Experimental Fluid Mechanics in Göttingen, Germany, and a professor of physics at the Australian National University, Canberra. He also held a position as a research scientist at the Aeronautical Research Laboratories in Melbourne. Dr. Hornung holds a bachelor's in mechanical engineering and a master's in engineering science from the University of Melbourne and a Ph.D. in aeronautics from Imperial College at the University of London. He is experienced in aerospace engineering, physics, thermodynamics, and aerodynamics.

Kathleen C. Howell is a professor at the Purdue University School of Aeronautics and Astronautics, where she has taught for 21 years. She holds a B.S. in aerospace engineering from Iowa State University, an M.S. in aeronautical and astronautical engineering from Stanford University, and a Ph.D. in aeronautical and astronautical sciences from Stanford University. Her research efforts focus on mission planning, spacecraft navigation, and maneuver requirements for vehicles in interplanetary space or in the neighborhood of Earth. Dr. Howell previously held positions with the

Proctor & Gamble Manufacturing Company and the Jet Propulsion Laboratory in Pasadena, California, and was a USAF/ASEE summer faculty fellow at the Air Force Rocket Propulsion Laboratory, where she researched auxiliary propulsion systems for attitude control of large, flexible space structures. She has experience in aerospace engineering, orbital mechanics, mission design, and spacecraft trajectory and attitude control.

Eric J. Jumper is a professor in the Aerospace and Mechanical Engineering Department at the University of Notre Dame, where he is also a member of the Center for Flow Physics and Control and directs the Aero-Optics Lab in the University of Notre Dame's Hessert Laboratory for Aerospace Research. His research includes work on aero-optics, turbomachines and turbofans, and aircraft wake dynamics. He has also taught at both the United States Air Force Academy in Colorado Springs, Colorado, and the Air Force Institute of Technology in Dayton, Ohio. In addition to his academic appointments, Dr. Jumper has worked as a research aerodynamicist and as chief of the Laser Devices Division at the Air Force Weapons Laboratory. He holds a B.S. from the University of New Mexico, an M.S. in mechanical engineering from the University of Wyoming, and a Ph.D. in gas dynamics and laser physics from the Air Force Institute of Technology. His expertise is in military acquisition and procurement, aerospace engineering, space science, government technical program management, physics, thermodynamics, propulsion and combustion, orbital mechanics, aerodynamics, reentry heating and thermal protection materials, surface chemistry, and aero-optics.

Ira F. Kuhn, Jr., is the founder, CEO, and president of Directed Technologies, Inc. (DTI), an R&D and science consulting firm specializing in projects relating to energy and national defense. DTI has completed projects involving the National Aerospace Plane, supersonic transports, and building a hydrogen infrastructure. Prior to founding DTI, Mr. Kuhn was a founder and vice president of Science and Analysis at B-K Dynamics, Inc., where he led a number of projects, including work with high-altitude unmanned surveillance platforms and long-range ramjets and antiaircraft missiles. He also worked as a senior scientist at Booz-Allen Applied Research and was a member of the National Security Advisory Panel for the Director of Central Intelligence and is now a member of the Army Science Board. Mr. Kuhn holds a B.S. in physics and an M.S. in industrial administration from Carnegie Mellon University. He has experience with military operations, acquisition and procurement, aerospace engineering, physics, systems engineering, and aerodynamics.

Andrew J. Meade is an associate professor of mechanical engineering at Rice University. His research focuses on machine learning and its application to experimental and computational mechanics. During 18 years of collaboration with NASA and the DoD, he has worked on projects involving aerodynamics, three-dimensional boundary layer separation, mesh-free finite element analysis, dynamic systems, multidisciplinary optimization, parallel and distributed computing, and heat transfer. Dr. Meade holds a B.S. from Rice University, an M.S. and a Ph.D. from the University of California at Berkeley in mechanical engineering, and is an American Institute of Aeronautics and Astronautics associate fellow. As the American Society of Mechanical Engineers and National Society of Black Engineers faculty advisor and the department's undergraduate advisor, he was active in the recruitment and mentoring of engineering students. He has experience in aerodynamics, multidisciplinary optimization, numerical and experimental methods, and aerospace workforce issues.

Carl J. Meade (U.S. Air Force, retired) is currently a portfolio manager in the Advanced Development Programs (Skunk Works) organization at the Lockheed Martin Aeronautics Company, where he previously held the position of X-33 program director. While in military service, Colonel Meade was the deputy division chief of the Crew and Thermal Systems Division as well as an Astronaut Office branch chief at NASA Johnson Space Center. Colonel Meade is a former F-16 aircraft commander, Air Force experimental test pilot, and NASA astronaut. He holds a B.S. in electrical engineering from the University of Texas, Austin, and an M.S. in electrical engineering from the California Institute of Technology. He is also a registered professional engineer in the state of Texas. He has experience in military operations, acquisition and procurement, aerospace engineering, space science, materials science and engineering, government technical program management, strategic planning, systems engineering, orbital mechanics, system verification, software verification, and flight-test planning and operations.

Neil E. Paton (NAE) is the chief technology officer and chairman of the Technology Advisory Board for Liquidmetal Technologies, where he has worked since March 2002. Prior to joining Liquidmetal, he served for 12 years as vice president of technology for the Howmet Corporation and as the president of Howmet Research Corporation, which developed products, processes, and materials for gas turbines. He also worked in materials development and advanced engineering for 20 years at Rockwell International, where he was involved in numerous programs, including the space shuttle program and the National Aerospace Plane (NASP) program. He holds B.S. and M.S. degrees in mechanical engineering from the University of Auckland, New Zealand, and a Ph.D. in materials science from Massachusetts Institute of Technology. His experience is in aerospace engineering, space science, materials science and engineering, government technical program management, propulsion engineering, and manufacturing operations.

Ronald F. Paulson is the corporate vice president of engineering at the Lockheed Martin Corporation. He has also served as vice president of remote sensing and space science at Missiles and Space, where he was responsible for 18 civil, national, and commercial space vehicle programs, and as vice president of technical operations with the Space Systems Company, where he was responsible for program performance, engineering, technology, and operations. Dr. Paulson holds B.S., M.S.M.E., and Ph.D. degrees in mechanical engineering from the University of Minnesota and an M.S. in management/business from Stanford University. His experience is in aerospace engineering, space science, materials science and engineering, government technical program management, physics, thermodynamics, strategic planning, systems engineering, aerodynamics, and satellite development and manufacturing.

Fred E. Saalfeld retired in 2002 as the executive director and technical director of the Office of Naval Research (ONR). This was the last in a series of positions over a 40-year career at ONR and the Naval Research Laboratory (NRL), where he spent time as superintendent of the Chemistry Division, director of the ONR Research Department, and Deputy Chief of Naval Research. Dr. Saalfeld received a B.S. degree in chemistry, physics, and mathematics from Southeast Missouri State University and M.S. and Ph.D. degrees in physical chemistry from Iowa State University. He has experience in materials science and engineering, government technical program management, strategic planning, and military research planning and policy.

Donna L. Shirley is director of the Science Fiction Museum in Seattle. She is also president of Managing Creativity, a consulting and training firm. She recently retired as assistant dean of engineering for Advanced Program Development and as an instructor in aerospace mechanical engineering at the University of Oklahoma. These positions followed a 33-year career with the Jet Propulsion Laboratory in Pasadena, California, where she served in a number of capacities, including managing the NASA Mars Exploration Program and overseeing the Pathfinder and Mars Global Surveyor missions and the Sojourner Rover program. Professor Shirley holds a B.A. in journalism and a B.S. in aerospace engineering from the University of Oklahoma and an M.S. in aerospace engineering from the University of Southern California. She has experience in aerospace engineering, space science, government technical program management, and systems engineering.

Peter Staudhammer (NAE) is an independent consultant and the retired vice president of science and technology at TRW, Inc. He served as the chief technology officer at TRW for 10 years, where he was responsible for the overall health and direction of the company's technical affairs, establishing technical leadership in space, electronics, and information systems and standardizing TRW's global product development and introduction processes. He also held positions as the chief engineer of the Apollo lunar descent engine and the Viking biology instrument programs and as director of the Physical Research Laboratory. He has been the chair of the NAE Program Committee since his induction into the Academy in 1996. Dr. Staudhammer holds a B.S. in electrical engineering and M.S. and Ph.D. degrees in chemical engineering from the University of California at Los Angeles. His expertise is in aerospace engineering, space science, thermodynamics, propulsion engineering, systems engineering, orbital mechanics, and engineering management.

B

Guest Speaker Presentations to the Committee

MEETING 1, WASHINGTON, D.C., AUGUST 5-6, 2003

OSTP Perspectives of NAI
Bill Jeffrey
White House Office of Science and Technology Policy

The National Aerospace Initiative—A Synergistic Partnership between NASA and DoD
John Rogacki
Headquarters, National Aeronautics and Space Administration

NAI and the U.S. Army
Billy Walker
U.S. Army Aviation and Missile Command

DARPA Programs for the NAI Objective
Art Morrish
Defense Advanced Research Projects Agency

Air Force CONOPS Champion Process and the Linkage to Science and Technology
Mike Snodgrass
Headquarters, U.S. Air Force

National Aerospace Initiative
Ron Sega
Director of Defense Research and Engineering

MEETING 2, DAYTON, OHIO, SEPTEMBER 3-5, 2003

Hypersonic Technology Status and Development Roadmap
Kevin Bowcutt
The Boeing Company

Hypersonics
Tom Corbett
Headquarters Air Combat Command

Hypersonics—A Lockheed Martin Perspective
Paul Hagseth
Lockheed Martin Corporation

AFRL and NAI Perspectives to AFSTB Committee on NAI
Jack Blackhurst
Air Force Research Laboratory

National Aerospace Initiative Overview
Jess Sponable
Air Force Research Laboratory

The Role of Turbine Engines in NAI
Timothy Lewis
Air Force Research Laboratory

AFRL Hypersonic Propulsion Presentation to the Air Force Science and Technology Board
Robert Mercier and Thomas Jackson
Air Force Research Laboratory

Power and Thermal Technologies for NAI
JoAnn Erno
Air Force Research Laboratory

Intersection of Air Force Propulsion Science and Technology and the NAI
Drew DeGeorge
Air Force Research Laboratory

National Aerospace Initiative: Materials and Processes
Daniel Evans
Air Force Research Laboratory

National Aerospace Initiative Payloads
Chuck Kastenholz
Air Force Research Laboratory

AFRL NAI Summary
Jack Blackhurst
Air Force Research Laboratory

NASA Hypersonics
Charles McClinton
NASA Langley Research Center

U.S. Hypersonic Ground Test and Evaluation Capabilities
Dan Marren
Arnold Engineering Development Center

National Aerospace Initiative USSTRATCOM Perspective
William Shelton
United States Strategic Command

MEETING 3, COLORADO SPRINGS, COLORADO, OCTOBER 7-9, 2003

Space Transportation Strategy
Gary Martin
NASA

Key Army Capabilities Where Hypersonic Technologies May Apply
Tim Coffin
Headquarters, Department of the Army

NGLT Briefing to the Air Force Science and Technology Board
Garry Lyles
NASA Marshall Space Flight Center

Requirements and Facilities for Ground Test of Full-Scale Scramjet Engines at Duplicated Flight Conditions from Mach 8 to 15
Michael Holden
Calspan-University of Buffalo Research Center

Synthesizing Numerical Analysis Data to Design and Develop Hypersonic Engines and Airframes
Graham Candler
University of Minnesota

HyFly Briefing to National Academy of Sciences
Gil Graff
Office of Naval Research

NAI Overview
Ron Sega
Director of Defense Research & Engineering

Rocketdyne Propulsion Efforts Related to NAI
John Vilja
The Boeing Company

AFSPC Perspective on the National Aerospace Initiative (NAI)
Pam Stewart
Air Force Space Command

Prompt Global Strike
Tom Ross
Air Force Space Command

Operational Concept for the Land Based Strategic Deterrent
Rod Peoples
Air Force Space Command

ORS Mission Need
Phil Pepperl
Air Force Space Command

ORS Operations Concept
Walt Koozin
Air Force Space Command

ORS Analysis of Alternatives
Pam Stewart
Air Force Space Command

Technology Assumptions
Robert Hickman
The Aerospace Corporation
Representing the Space and Missile Systems Center

Military Utility Assessment
Greg Keether
Air Force Space Command

Candidate Space Architectures
Robert Hickman
The Aerospace Corporation
Representing the Space and Missile Systems Center

Launch Vehicle Alternatives
Robert Hickman
The Aerospace Corporation
Representing the Space and Missile Systems Center

Cost Effectiveness Methodology
Dave Boudreaux
Air Force Space Command

ORS Ahead
Pam Stewart
Air Force Space Command

Applicability of Demo Programs to ORS
Gus Hernandez
Air Force Space Command

Perspectives on Hypersonics
Pete Worden
Space and Missile Systems Center

NAI and the U.S. Army
Billy Walker
U.S. Army Aviation and Missile Command

Achieving Routine Affordable Responsive Access to Space
Robert Hickman
The Aerospace Corporation

MEETING 4, WASHINGTON, D.C., NOVEMBER 10-11, 2003

NAI Introduction and Summary
Ron Sega
Director of Defense Research and Engineering

NAI High Speed/Hypersonics Pillar
Glenn Liston
NAI Executive Office

NAI Space Access Pillar
Uwe Hueter
NAI Executive Office

Threat Briefing
Sean Sedell and Bob Merkel
Central Intelligence Agency

C

National Aerospace Initiative Hypersonics Programs and Technologies

NATIONAL AERONAUTICS AND SPACE ADMINISTRATION

X-43A Hyper-X

The first demonstrator vehicle in NASA's "Hyper-X" series of experimental hypersonic ground and flight test vehicles, the X-43A will demonstrate "air-breathing" engine technologies for future hypersonic aircraft and/or reusable space launch vehicles, achieving speeds above Mach 5, or five times the speed of sound [Figure C-1]. The X-43A is intended to dramatically increase payload capacity or reduce vehicle size. A successful flight could mark the first time a nonrocket, air-breathing supersonic-combustion ramjet—or "scramjet"—engine has powered a vehicle in flight at hypersonic speeds. The next flight test of the X-43A is scheduled for January 2004. (NASA, 2003a)

X43-B

The X-43B will be the first reusable vehicle in the X-43 series of X-planes. Not requiring a booster, it will be air-launched from NASA's new B-52H aircraft, in a manner similar to the X-15, and then flown to Mach 7 under its own power. It is an ambitious undertaking and will go a long way toward advancing technology readiness levels for future hypersonic aircraft and air-breathing space access vehicles.

Two different engine systems are being considered for the X-43B. The first is a rocket based combined cycle (RBCC) engine, which is a scramjet with rockets imbedded in the internal flow-path. With this engine approach, the rockets are operated initially, up to a flight speed of Mach 3 or 4. The rockets are then turned off, and the engine is operated in ram/scramjet mode to Mach 7-8 for hydrocarbon fuel, or Mach 12-15 for hydrogen fuel. The second is a turbine based combination cycle (TBCC) system, which uses separate low and high-speed engines. Hydrocarbon-fueled turbo-jet or turbo-ramjet engines are used for flight up to Mach 4, and then separate ram/scramjet engines are used for flight up to Mach 7-8 (hydrocarbon fuel) or Mach 12-15 (hydrogen fuel). (Orton, 2002)

FIGURE C-1 X-43A hypersonic experimental vehicle—artist concept in flight. SOURCE: NASA, 1999.

X-43C Hypersonic Demonstration Vehicle

Now in development, the X-43C is expected to accelerate to a maximum potential speed of about 5,000 mph, and could undergo flight-testing as early as 2008. NASA will develop, test and fly the Hyper-X series over the next two decades to support development of future-generation reusable launch vehicles and improved access to space. (NASA, 2002a)

The X-43C demonstrator, powered by a scramjet engine developed by the U.S. Air Force, is now in development. The X-43C is expected to accelerate from Mach 5 to Mach 7, reaching a maximum potential speed of about 5,000 mph. NASA will begin flight-testing the X-43C in 2008. (NASA, 2002b)

NASA's Advanced Space Transportation Program (ASTP) Hypersonic Investment Area

NASA Marshall's Advanced Space Transportation Program (ASTP) is investing in hypersonics as part of a new national hypersonics strategy formulated by NASA, the Air Force, Army, Navy and Defense Advanced Research Projects Agency. (NASA, 2003b)

Rocket-Based Combined Cycle Engines

Rocket-Based, Combined-Cycle (RBCC) engines are currently being explored as advanced propulsion for a variety of space launch, cruise aircraft and unmanned systems. This flexible propulsion

> system can be customized to meet the unique needs of both commercial and military applications. RBCCs can utilize either storable or cryogenic propellants dependent on system requirements. RBCC powered systems can provide significant advantages in range, mission time, weight, payload, load-out, mission profile flexibility, and cost over competing conventional propulsion solutions. The Boeing RBCC can be scaled, so that development and demonstration of the basic core flowpath will support a broad spectrum of postulated applications.
>
> The Boeing RBCC will operate in air-augmented rocket, ramjet, scramjet and rocket propulsive modes to deliver high average Isp at required thrust levels for the earth to orbit mission. Fuel injection locations are varied during the flight to accommodate the unique needs of each operating mode. (Boeing, 2003)

Turbine-Based Combined Cycle Engines

> NASA's Advanced Space Transportation Program has tasked Glenn Research Center (GRC) to lead the high Mach turbine propulsion and Turbine-Based Combined Cycle (TBCC) efforts. In response to this request, GRC has developed the RTA Project to develop and demonstrate high Mach turbine propulsion and Turbine-Based Combined Cycle (TBCC) propulsion for Space Access. (NASA, 2003c)
>
> The Turbine-Based Combined Cycle (TBCC) engine project seeks to deliver a Mach 4+ hypersonic propulsion system in this decade. (NASA, 2003d)

U.S. AIR FORCE

HyTech Program

This Air Force, NASA, and Pratt & Whitney hypersonic missile program is being developed using hydrocarbon fuel. Most programs have used only hydrogen as a fuel for scramjets because of the brief time for combustion (~1 ms). The Hypersonic Technology (HyTech) Program instead uses conventional jet fuel (JP-7). The heat generated in the walls of the combustion chamber is used to crack the fuel into lighter and more volatile elements. These elements then generate positive thrust at Mach 6 and higher when they enter the supersonic flow and burn. Future program goals will be to have engines useful for both missiles and reusable vehicles (Hobbyspace, 2003; Jackson, 2003).

> Pratt & Whitney (P&W) Space Propulsion, teamed with U.S. Air Force researchers under the Hypersonic Technology (HyTech) Program, has completed testing of a revolutionary scramjet engine. The ground demonstration engine number one (GDE-1), which weighs less than 150 pounds, was tested at speeds of Mach 4.5 and Mach 6.5 in hypersonic ground test facilities. GDE-1 was the world's first flight-weight, hydrocarbon-fueled scramjet engine, and used standard JP-7 fuel to both cool engine hardware and fuel the engine's combustor. (Pratt & Whitney, 2003)
>
> The objective of the HyTech program is to demonstrate the operability, performance, and structural durability of a liquid hydrocarbon (jet fuel) supersonic combustion ramjet (Scramjet). The near term application of this technology is a long range hypersonic cruise missile that is logistically supportable in a combat environment and can defeat time-sensitive targets and hard and deeply buried targets. In the far term, the scramjet technology enables a Mach 8-10 strike/reconnaissance aircraft and affordable, on-demand access to space with aircraft like operations. (AFRL, 2003)

Integrated High-Performance Turbine Engine Technology Program

> The Integrated High Performance Turbine Engine Technology (IHPTET) program, started in 1988, has an aggressive technology development plan to leapfrog technical barriers and deliver twice the propulsion capability of today's systems by around the turn of the century. Unprecedented teaming

of the Army, Navy, Air Force, NASA, DARPA and industry, in each of the technology areas, is underway. The main focus of these 'Technology Teams in Action' is to advance military aircraft superiority through high performance, affordable, robust turbine engines. (AFRL, 2003)

The Integrated High Performance Turbine Engine Technology (IHPTET) program started in 1988, with a goal of doubling the propulsion capability by the year 2005. To meet this goal the Army, Navy, Air Force, NASA, DARPA, and industry united to develop high performance, affordable, robust turbine engines.

The development of new materials and advanced structural designs is crucial to IHPTET's success and future robust engine systems with stronger and more durable components. With this emphasis on new materials development, many breakthroughs in high temperature materials and processes owe their success to the IHPTET program. Furthermore, many of these materials form the foundation for more efficient rocket engines being examined under a similar program (Integrated High Payoff Rocket Propulsion Technology—IHPRPT). (IHPTET, 2003)

Single-Engine Demonstrator

The Single-Engine Demonstrator (SED) program will demonstrate scramjet engine and vehicle technologies in a relevant flight environment, making them ready for transition to weapon system development. The primary program objective is to validate performance of the fixed-geometry hypersonic technology (HyTech) scramjet engine and the integration of that engine into an airframe based on the ARRMD vehicle design. The planned mission profile is to have the demonstrator carried by a B-52 aircraft to an altitude of about 35,000 feet and released (Pratt & Whitney, 2004). Initially propelled by a solid rocket booster, the recoverable scramjet demonstrator takeover will occur at approximately Mach 4.5, where it then will accelerate to flight speed between Mach 6.0 and 7.0+. This flight profile will expose the demonstrator to both high dynamic pressure during the acceleration phase and low dynamic pressure in the cruise phase. Five to eight flight tests are planned, but additional flights are possible for testing related technologies such as thermal protection systems, guidance, high speed dispense, and so on (McDaniel, 2004).

Atmospheric Interceptor Technology Program

The United States Air Force atmospheric interceptor technology (AIT) program is developing, integrating and demonstrating lightweight launch vehicle technologies within the atmosphere to support ballistic missile technology. The Air Force has a requirement to develop target vehicles that have the ability to simulate a realistic incoming ballistic missile trajectory. This capability will be used to evaluate the performance and utility of currently existing radar on the West Coast of the United States and provide an experimental platform for evaluation of future technology. (U.S. Air Force, 1999)

AIT aims to improve the capability of ballistic missile defense systems that operate within the earth's atmosphere, such as theater high altitude area defense and the Navy area defense system. AIT also seeks to double average velocities, resulting in increases of defended areas by about a factor of three. These performance improvements are relevant to cruise missile defense as well as ballistic missile defense.

The AIT program is composed of four critical technology areas: a strapdown seeker design, a cooled window, a solid divert-and-attitude control system and a composite material airframe. (Granone & Davis, 2000)

U.S. NAVY

RATTLRS Program

Revolutionary Approach to Time-critical Long Range Strike (RATTLRS) Flight Demonstration Project. Sources/respondents should have a demonstrated capability to successfully design, fabricate, integrate and flight test a tactical strike-weapon-class supersonic air vehicle. This notice requests information on project execution plans and identification of the requisite technologies to ascertaining the feasibility and risks associated with such a flight demonstration project. The total projected technology development and flight demonstration cost is approximately $50M. As an applied research Science and Technology demonstration, RATTLRS will culminate in flight demonstrations of an integrated air vehicle, powered only by a supersonic turbine engine, in a size/shape/weight configuration that is traceable to a tactical weapon system. The flight demonstration vehicle is expected to accelerate from a subsonic air launch condition to a minimum of Mach 3 using only turbine propulsion. For RFI response purposes, assume first demonstration flight prior to 36 months after project initiation, and all flight demonstrations completed by 45 months after project initiation. The flight vehicle to be demonstrated within RATTLRS should be a demonstration air vehicle only, having traceability to one of the following potential weapon systems: An air-launched (compatible with the F/A-18 launch platform) medium range weapon with a maximum vehicle weight of 1800 lbs including 500 lbs payload, or a ship-launched (Vertical Launch System (VLS) compatible) long-range weapon with a maximum weight of 3400 lbs including 750 lbs payload (includes vehicle and any booster required for VLS launch), or one vehicle compatible to both launch options. (U.S. Navy, 2003)

U.S. ARMY

Scramfire

Scramfire is a 120-mm powered munition "that accelerates throughout flight to the target and offers increased velocity at the target for direct fire weapons or increased range for indirect fire" (CAS, 2003). The scramjet-powered munition (see Figure C-2) demonstrated structural integrity by inflight x-ray and had a self-sustaining thrust that enabled a flight velocity of 8,100 ft/sec over 240 ft flight (Sega, 2003).

Hypersonic Interceptor Missile Scramjet Program

Provide optimized design for a Mach 10-12 H_2 fueled scramjet powered interceptor missile to be used for defense against cruise missiles and for Army long range attack operations (long range

FIGURE C-2 Scramjet powered munitions. SOURCE: Sega, 2003.

strike). Work is to be performed in collaboration with test program providing full scale shock tunnel test data at fully duplicated flight conditions, and, with a team of University experts performing subscale experiments. (Kennedy and Nash, 2003)

DEFENSE ADVANCED RESEARCH PROJECTS AGENCY

DARPA/Navy HyFly Program

The Hypersonics Flight Demonstration (HyFly) program will develop and demonstrate advanced technologies for hypersonic flight. Flight-testing will be initiated early in the program and progress from relatively simple and low-risk tests through the demonstration of an increasingly more difficult set of objectives. The ultimate goals of the program are to demonstrate a vehicle range of 600 nautical miles with a block speed of 4,400 ft per sec, maximum sustainable cruise speed in excess of Mach 6, and the ability to deploy a simulated or surrogate submunition. Technical challenges include the scramjet propulsion system, lightweight, high-temperature materials for both aerodynamic and propulsion structures, and guidance and control in the hypersonic flight regime. Recently demonstrated performance in ground testing of the dual combustion ram-jet engine coupled with advances in high temperature, lightweight aerospace materials are enabling technologies for this program. The program will pursue a dual approach. The core program will focus on development and demonstration of capabilities requisite for an operational weapon. A separate effort will be performed in parallel to demonstrate advanced propulsion technologies and develop low-cost test techniques. DARPA and the Navy established a joint program to pursue areas of the hypersonics program that would be relevant to maritime applications. (DARPA, 2003a)

DARPA/USAF FALCON Program

The FALCON program objectives are to develop and demonstrate technologies that will enable both near-term and far-term capability to execute time-critical, global reach missions. Near-term capability will be accomplished via development of a rocket boosted, expendable munitions delivery system that delivers its payload to the target by executing unpowered boost-glide maneuvers at hypersonic speed. This concept called the Common Aero Vehicle (CAV) would be capable of delivering up to 1,000 pounds of munitions to a target 3,000 nautical miles downrange. An Operational Responsive Spacelift (ORS) booster vehicle will place CAV at the required altitude and velocity. The FALCON program will develop a low cost rocket booster to meet these requirements and demonstrate this capability in a series of flight tests culminating with the launch of an operable CAV-like payload. Far-term capability is envisioned to entail a reusable, hypersonic aircraft capable of delivering 12,000 pounds of payload to a target 9,000 nautical miles from CONUS in less than two hours. Many of the technologies required by CAV are also applicable to this vision vehicle concept such as high lift-to-drag technologies, high temperature materials, thermal protection systems, and periodic guidance, navigation, and control. Initiated under the Space Vehicle Technologies program, and leveraging technology developed under the Hypersonics program, FALCON will build on these technologies to address the implications of powered hypersonic flight and reusability required to enable this far-term capability. The FALCON program addresses many high priority mission areas and applications such as global presence, space control, and space lift. (DARPA, 2003b)

REFERENCES

Published

AFRL (Air Force Research Laboratory). 2003. Technology Programs. Available at http://www.pr.afrl.af.mil/technology.htm. Accessed on December 23, 2003.

Boeing. 2003. Rocket Based Combined-Cycled Engine. Available at http://www.boeing.com/defense-space/space/propul/rbcc.html. Accessed on December 23, 2003.

CAS (Committee on Armed Services, House of Representatives). 2003. National Defense Authorization Act for Fiscal Year 2004. Report of the Committee on Armed Services, House of Representatives on H.R. 1588. Available at http://armedservices.house.gov/billsandreports/108thcongress/H1588RAM.pdf. Accessed on January 13, 2004.

DARPA (Defense Advanced Research Projects Agency). 2003a. Force Application and Launch from CONUS (FALCON). Available at http://www.darpa.mil/tto/programs/falcon.html. Accessed on December 23, 2003.

DARPA. 2003b. Hypersonic Flight (HyFly). Available at http://www.darpa.mil/tto/programs/hyfly.html. Accessed on December 23, 2003.

Granone, J.F., and W.A. Davis. 2000. Anti-missile Systems Should Not Be Rushed: Emerging Technologies Offering New Ways to Defeat Ballistic-Missile Attacks. Available at http://www.nationaldefensemagazine.org/article.cfm?Id=216. Accessed on December 23, 2003.

Hobbyspace. 2003. Launch & Propulsion Systems, Scramjet & Other Airbreathing Propulsion. Available at http://www.hobbyspace.com/Links/LaunchPropulsion2.html. Accessed on December 23, 2003.

IHPTET (Integrated High Performance Turbine Engine Technology). 2003. IHPTET Virtual Library: A Word About IHPTET. Available at http://namis.alionscience.com/virtual/Ihptet/about_IHPTET.htm. Accessed on December 23, 2003

Jackson, T. 2003. Hypersonic technology scramjet engine reaches major milestone. AFRL Horizons. Available at http://www.afrlhorizons.com/Briefs/Dec01/PR0102.html. Accessed on December 23, 2003.

Kennedy, K., and S. Nash. 2003. Computational Hypersonic Scramjet Technology Enhancements for Long-Range Interceptor Missile. DoD High-Performance Computing Modernization Program. FY2004 Challenge Project Abstracts. U.S. Army, Army Aviation and Missile Command. Available at http://www.hpcmo.hpc.mil/Htdocs/Challenge/FY04/4.html. Accessed on December 23, 2003.

NASA. 1999. NASA Dryden Flight Research Center Photo Collection, Photo ED99-45243-01. Available at http://wwwdfrc.nasa.gov/gallery/photo/index.html. Accessed on December 23, 2003.

NASA. 2002a. NASA Developing Hypersonic Technologies; Flight Vehicles Only Decades Away. NASA Photo Release No. 02-182, July 22. Available at http://www1.msfc.nasa.gov/NEWSROOM/news/photos/2002/photos02-182.html. Accessed on December 23, 2003.

NASA. 2002b. NASA Developing Hypersonic Technologies; Flight Vehicles Only Decades Away. NASA News Release No. 02-182, July 22. Available at http://www1.msfc.nasa.gov/NEWSROOM/news/releases/2002/02-182.html. Accessed on December 23, 2003.

NASA. 2003a. Next Generation Launch Technology (NGLT) News–X43-A Website. Available at http://www.ngltnews.com/x43a.html. Accessed on December 23, 2003.

NASA. 2003b. Hypersonics. Available at http://www.spacetransportation.com/ast/hypersonics.html. Accessed on December 23, 2003.

NASA 2003c. Turbine-Based Combined Cycle (TBCC) Revolutionary Turbine Accelerator (RTA) Project. Available at http://tbcc.grc.nasa.gov/index.shtml. Accessed on December 23, 2003.

NASA. 2003d. Next Generation Launch Technology (NGLT) News–TBCC/RTA. Available at http://www.ngltnews.com/tbcc.html. Accessed on December 23, 2003.

Orton, G.F. 2002. Air-Breathing Hypersonics Research at Boeing Phantom Works. Abstract available at http://hypersonic2002.aaaf.asso.fr/papers/17_5251.pdf. Accessed on December 23, 2003.

Pratt & Whitney. 2003. Flight-Type Scramjet Completes Historic Test Series. Press release. Available at http://www.pw.utc.com/pr_071403.asp. Accessed on December 23, 2003.

Pratt & Whitney. 2004. Air Force Selects Pratt & Whitney and Boeing Team for the Scramjet Flight Demonstrator Program. Available at http://www.pratt-whitney.com/pr_011304.asp. Accessed on March 25, 2004.

U.S. Air Force. 1999. Air Force Successfully Launches ait-2 Rocket. USAF Press Release. Available at http://www.losangeles.af.mil/SMC/PA/Releases/1999/nr99091501.htm. Accessed on December 23, 2003.

U.S. Navy. 2003. FEDBIZOPS. Special Notices. April 2003—China Lake and Point Mugu. Available at https://contracts.nawcwd.navy.mil/cbdsp_p04.htm. Accessed on December 23, 2003.

Unpublished

McDaniel, J. 2004. Single-Engine Demonstrator. Jay McDaniel response to a committee question, received by staff on January 14.

Sega, R. 2003. National Aerospace Initiative: NRC Briefing. Briefing by Ron Sega, DDR&E, Office of the Secretary of Defense, to the Committee on the National Aerospace Initiative. August 6.

D

National Aerospace Initiative Access-to-Space Programs

SPACE ACCESS TECHNOLOGY PROGRAMS

Next-Generation Launch Technology

NASA's Next-Generation Launch Technology (NGLT) program has as an objective the development and demonstration of high leverage and cross-cutting technologies to enable safe, reliable, and affordable space launch systems capabilities. Currently funded at over $500 million per year, the effort is directed to shuttle life extension and expendable launch upgrades in the near term, exploration launch in the medium term, and new launch capabilities in the far term.

The NGLT program seeks to develop and mature innovative technologies with new research in the areas of propulsion, structures, vehicle systems, and ground and flight operations. The program is currently pursuing development of a reusable liquid oxygen/liquid kerosene rocket booster engine; hypersonic, air-breathing propulsion and airframe systems; crosscutting launch vehicle system technologies intended to support a broad variety of launch and flight vehicle architectures; and analysis efforts to guide program investment.

In 2004, the program will decide whether to proceed with a next-generation launch vehicle risk-mitigation phase, which includes research and testing of large-scale tanks, structures, and engines. The following list is a representative list of the NASA NGLT programs:

- X-43A
- X-43C
- Revolutionary turbine accelerator
- Rocket-based combined cycle

Air Force Research Laboratory Programs

The Air Force Research Laboratory (AFRL) is the technology development arm of the Air Force for hypersonics, space access, and spaceborne systems. Current NAI-specific and -related funding within AFRL is over $200 million, with space access receiving about $100 million of support. NAI plus-ups could add $25 million to the Space Access numbers for FY 2004. AFRL programs support upgrades to current expendable launch systems and new capabilities in rapid launch, reusable launch vehicles, rapid turnaround operations, and so on, for support of the future mission needs of the Air Force commands.

The activities in Space Access pursued by AFRL include propulsion, thermal protection systems, aerothermodynamics of high-speed vehicles, integrated vehicle health management systems, simulation-based research and development, and others. The following is a representative list of AFRL technology programs:

- Integrated high-payoff rocket propulsion technology (IHPRPT)
- Integrated powerhead demonstrators
- Reusable access to space-Technology (RAS-T)
- X-43B
- Force Application and Launch from CONUS (FALCON)
- Hydrocarbon scramjet development (one effort of HyTech)

SPACE ACCESS DEMONSTRATION PROGRAMS

There are significant NAI demonstration programs within DARPA.

Responsive Access, Small Cargo, Affordable Launch

Responsive Access, Small Cargo, Affordable Launch (RASCAL) is a three-phase demonstration program to provide a path toward a revolution in rapid access to space. RASCAL consists of a reusable aircraft as the launch platform and a two-stage expendable rocket vehicle for accelerating a 165-lb payload to orbital velocity. The RASCAL aircraft utilizes standard military jet engines augmented by mass injection precompressor cooling (MIPCC), which allows operation at higher velocities and altitudes. Figure D-1 illustrates the system.

While still in the atmosphere, the RASCAL aircraft accelerates to a high velocity and begins a steep climb. With MIPCC, the aircraft accelerates while climbing until the engines are shut down at 100,000 feet and then coasts to an altitude of 200,000 feet, where the expendable rockets are released. The aircraft then reenters the atmosphere and returns to its base of operations. An operational system would be capable of rapid turnaround.

Force Application and Launch from CONUS

DARPA and the Air Force are jointly sponsoring the FALCON program to develop technologies and demonstrate capabilities that will enable Prompt Global Strike from the continental United States. The small launch vehicle (SLV) that is part of FALCON (Figure D-2) will serve a twofold function in that it will also provide a low-cost responsive capability for launching satellites up to the 2,000-lb class into Sun-synchronous orbit.

For the access-to-space application, the SLV must be at least an order of magnitude more responsive than existing satellite launch systems and must have a low launch cost. The program will pursue development of an innovative concept possessing these attributes for flight demonstration in the 2007 time frame. The program will also seek to develop a unique CONOPS that will support and enable both the responsiveness and low-cost system objectives.

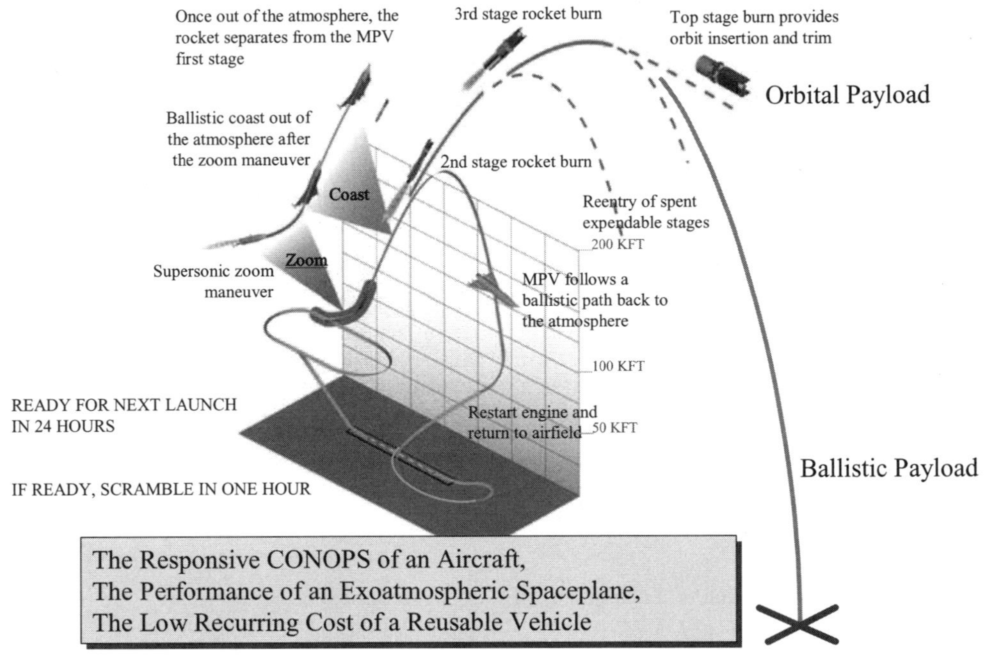

FIGURE D-1 RASCAL CONOPS. SOURCE: Morrish, 2003.

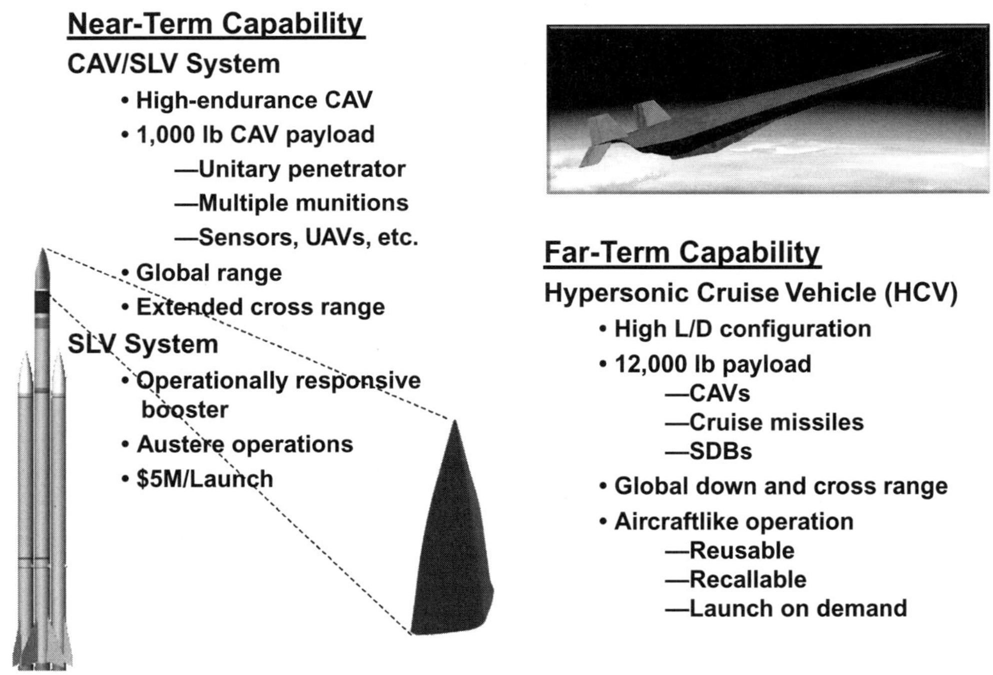

FIGURE D-2 FALCON approach. SOURCE: Morrish, 2003.

FUTURE SPACE ACCESS PROGRAMS WITHIN THE NATIONAL AEROSPACE INITIATIVE

Air Force

Operationally Responsive Spacelift

Currently, in the analysis-of-alternatives phase, the ORS program will develop a capability to launch, maneuver, service, and retrieve space payloads. The key element is responsiveness, with the long-term goal of hours vs. today's experience of weeks to months. The military spaceplane system architecture shown in Figure D-3 is a current concept of an operational system.

The access-to-space elements of the system are the space operations vehicle (SOV), the modular insertion stage (MIS), and the space maneuver vehicle (SMV). The SOV is believed to be the most difficult and will require a phased demonstration approach to mitigate risk in areas such as fast turn propulsion, thermal protection systems, metallic tank and insulation, power and actuation systems, and ground operations.

Small, Low-Cost Launch Vehicle

A possible outgrowth of the joint Air Force/DARPA FALCON program is a small, low-cost, expendable launch vehicle capable of rapid launch and available substantially before the military space plane (MSP).

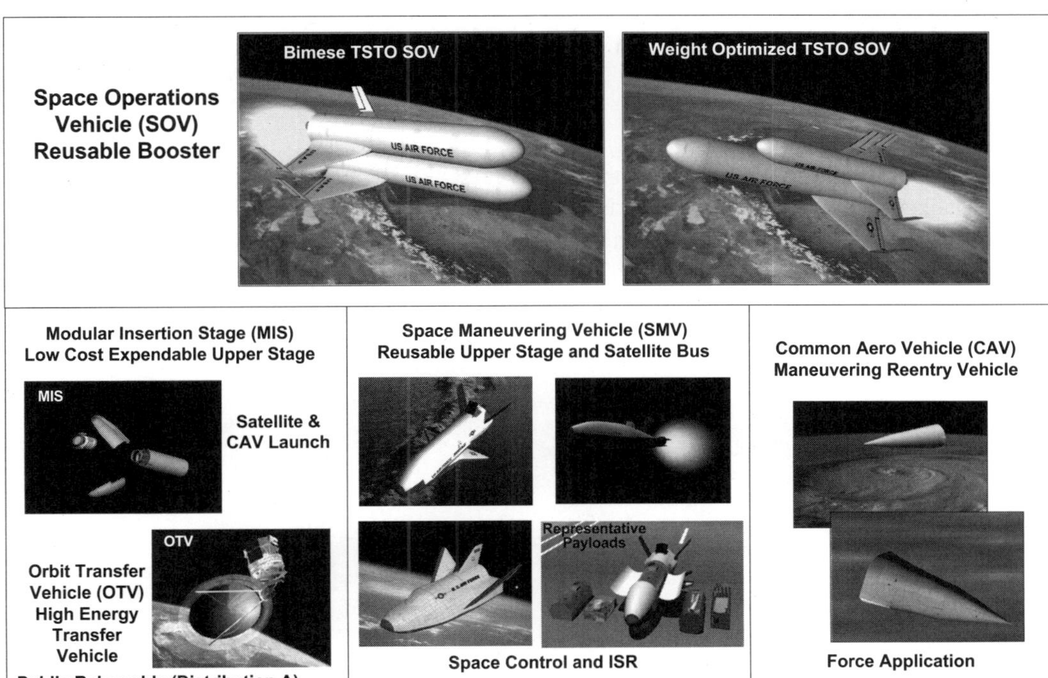

FIGURE D-3 Military spaceplane system architecture. SOURCE: Koozin, 2003.

X-42

The X-42 is envisioned to be a flying testbed for validating selected technologies developed since the late 1980s. In addition to being an environment simulator for airframe and flight subsystems technologies, NAI advertises that the X-42 will be used to evaluate improved technologies and approaches for fighterlike operability in a military space plane-type system. If designed properly, it can also be used as a payload integration testbed enabling flight testing of next-generation air-breathing propulsion technologies. As envisioned today, X-42 would be scalable to the second stage of a two-stage-to-orbit space plane. Scalability ensures that component technologies will be developed and flight-tested in environments relevant to the operational end system.

Upgrades to Current Expendable Launch Vehicles

Replacements for the Russian RD-180 booster engine and the Centaur and Inertial Upper Stage are under consideration.

National Aeronautics and Space Administration

Expendable Launch Vehicles

Upgrade of an existing expendable launch vehicle to carry the crew exploration vehicle into low Earth orbit for crew transfer to the International Space Station is a possibility. The crew exploration vehicle is currently being studied for possible implementation in this decade, first as a crew rescue vehicle and soon thereafter as a crew transfer vehicle. Like its predecessor, the orbital space plane, it is not anticipated that the crew exploration vehicle will fall under the umbrella of NAI.

Development of a heavy-lift (130 metric tons), expendable launch system for both crewed and uncrewed exploration of the solar system has gained support among some at NASA.

Reusable Launch Vehicle

Given appropriate funding and emphasis, the NGLT program could lead to a rocket-based reusable launch vehicle in the next decade as a replacement for the space shuttle, providing both uncrewed and crewed access to space. An overview of NASA's space transportation plan is shown in Figure D-4.

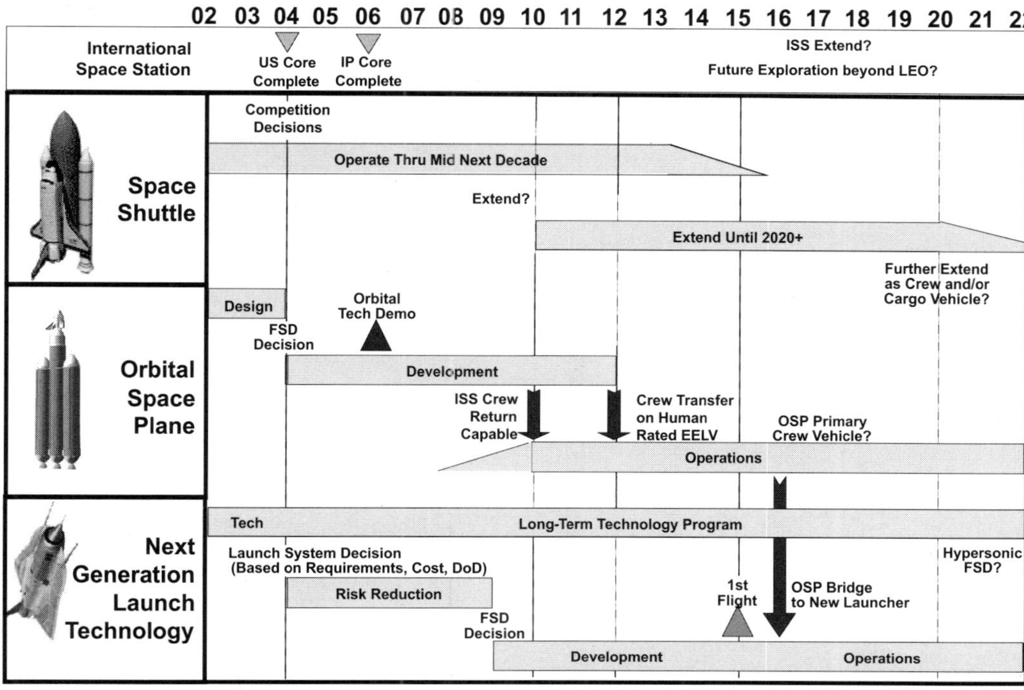

FIGURE D-4 Integrated Space Transportation Plan before the Presidential Vision announcement of January 14, 2004. SOURCE: Rogacki, 2003.

REFERENCES

Unpublished

Koozin, W. 2003. ORS Operations Concept. Briefing by Walt Koozin, Air Force Space Command, to the Committee on the National Aerospace Initiative. October 7.

Morrish, A. 2003. DARPA Programs for the National Aerospace Initiative Objective. Briefing by Art Morrish, DARPA, to the Committee on the National Aerospace Initiative. August 5.

Rogacki, J. 2003. The National Aerospace Initiative–A Synergistic Partnership between NASA and DoD. Briefing by John "Row" Rogacki, NASA, to the Committee on the National Aerospace Initiative. August 5.

E

Fuels Research

The heat sink capacity of a fuel—hydrogen, for example—is due to its sensible heat (i.e., $c_p \Delta T$) and is thus proportional to the maximum temperature the fuel can achieve and the magnitude of its specific heat. Hydrogen has a much higher specific heat than hydrocarbon fuels, and assuming that ΔT is the same, is considered to have a higher heat capacity than hydrocarbon fuels; however, less hydrogen mass is required to achieve a given amount of heat release, so the hydrogen fuel flow rate is lower than that of a hydrocarbon. A more relevant measure of heat sink potential is the ratio of specific heat to the heat of combustion, which brings hydrocarbon fuels more in line with hydrogen in terms of heat capacity (within a factor of 2 or 3). Thus, it is not so much the specific heat of hydrogen that gives it better heat capacity than hydrocarbon fuels but rather its ability to sustain higher temperatures and thus increase the ΔT part of its heat capacity. In the absence of other considerations, conventional hydrocarbon fuels cannot be driven to as high a temperature as hydrogen, which is capable of sustaining temperatures up to the operating temperature of the engine.

In the case of an endothermic fuel like JP-7, another source of heat capacity is made available by cracking the fuel, a process of breaking its long-chain hydrocarbon molecules into lighter molecules that absorb heat (an endothermic process). In an endothermic fuel, the heat sink capability of the fuel is made up of its sensible heat plus any net endothermic capacity derived from high fuel dissociation reactions. The key is to find a fuel that has endothermic capability without degrading its exothermic capability. Typically, hydrocarbon decomposition processes are accompanied by carbon formation, or coking. Coking tends to foul heat transfer surfaces, which is undesirable. Thus, there are two parts to calculating the upper limit of a hydrocarbon fuel's heat sink capability: the maximum temperature achievable without the system coking up, and the endothermic capacity of the cracking reactions that can occur.

Current U.S. state-of-the-art heat sink capability for an endothermic liquid hydrocarbon fuel under realistic conditions is roughly 1,500 Btu/lb, with a limit of 1300°F (Huang et al., 2002). Under operational conditions, a 1300°F temperature limit allows for a coking-limited lifetime of approximately 15 min. In this case, the total fuel heat sink capacity is a combination of 500 Btu/lb from endothermic reactions and roughly 1,000 Btu/lb from sensible heat. If the coking limit of a fuel could be raised by a few hundred degrees, its heat sink capability could be enhanced.

It is important to point out that the endothermic heat capacity of 500 Btu/lb is much less than the theoretical heat capacity achievable if the most desirable fuel reaction products could be obtained. Cracking JP-7 to 100 percent ethylene would absorb 1,500 Btu/lb, versus the 500 Btu/lb obtained from the product mix of methane, ethylene, ethane, etc. Thus, there is clearly room for increasing the endothermic capability of even our present candidate fuels. Because of the endothermic capability of our present hydrocarbon fuels, at the upper Mach number range for hypersonic air-breathing flight (approximately Mach >8), hydrogen is the fuel of choice; however, because of hydrogen's difficulty of storage and its limited volumetric energy density, hydrocarbon fuels are more practical for lower Mach number applications. All of the hydrocarbon-based vehicles presently being contemplated anticipate using JP-7. As previously discussed, a robust fuels research and development (R&D) program could lead to higher Mach number limits for hydrocarbon fuels. The Russian AJAX (concept) vehicle, for example, increases heat sink capability by an interesting variation on conventional endothermic fuels. In the AJAX concept, water is added to the fuel to achieve steam reforming. In essence, steam reforming is fuel + water $\rightarrow$ $CO + H_2$. This reaction absorbs (theoretically) 2,400 Btu/lb (versus 1,500 Btu/lb for cracking to ethylene). This is how the Russians theoretically obtain Mach 10 capability for AJAX, although the concept suffers reduced range owing to water consumption in the propulsion cycle.

GENESIS OF THE HYTECH/X-43C JP-7 FUEL "SYSTEM"

In the 1960s and early 1970s, the Air Force funded a considerable effort at Shell Research to develop endothermic methylcyclohexane (MCH), based on studies that identified the heat sink required for hydrocarbon-fueled hypersonic vehicles (Churchill et al., 1965; Lander and Nixon, 1971). Endothermic MCH required a supported platinum catalyst for dehydrogenation (to toluene and hydrogen) that was developed in pellet form (as is used industrially in fluidized beds for petroleum processing). After a long period of dormancy in the 1970s and 1980s, the work was restarted by Allied Signal (now Honeywell) and culminated in an expendable turbine-engine test, where the fuel was used to cool a hot air stream and then burned in an engine (Lipinski et al., 1992). There were two drawbacks to this technology: (1) regenerative cooling is best accomplished through a wall-mounted catalyst rather than pellets and (2) MCH is a relatively expensive specialty fuel.

Extensive discussions inside the Air Force Propulsion Directorate in the mid-1980s led to an effort to develop the endothermic potential of thermally/catalytically cracked liquid hydrocarbons using commercially available, wall-mounted zeolite catalysts (Spadaccini et al., 1993a). This task at the United Technologies Research Center was funded under program element 62203F/Project 3048 of the initial contract, F33615-87-C-2744, managed by Charlotte Eigel. This effort first reported the endothermic potential of JP-7 and JP-10, as well as JP-8 (Sobel and Spadaccini, 1994). The resulting endothermic hydrocarbon fuel capability contributed to selecting such fuels for the HyTech engine development program, which began in the mid-1990s (Spadaccini et al., 1993b). This separate (from HyTech) but coordinated fuel development effort has continued sporadically since HyTech funding became available. Recent tasks include extending the endothermic heat sink database to higher flow rate conditions and to other fuels (e.g., RP-1), as well as looking at relative combustion performance of the various alternative fuels (Huang and Sobel, 2002). The fuels research effort has also looked at the applicability of endothermic fuels to reusable aircraft and other applications (Lehrach et al., 1995).

A POSSIBLE FUELS RESEARCH PROGRAM

To provide some structure to a program that might be created by NAI, the committee consulted with Tim Edwards, a civilian fuels scientist at AFRL. Fuel is becoming the integrating factor of the complete hypersonic vehicle system.[1] It is difficult to give a reasonable estimate for a long-term endothermic fuels plan without bounding the problem, in that key drivers and constraints for fuel and fuel system development come from other nominally distinct disciplines—thermal management, combustion, and ground handling, to name a few. Thus, the fuel ultimately selected for a given hypersonic vehicle will probably be a compromise between cost, operability, and performance in a number of areas. NASA's vision vehicle for air-breathing space access is fueled by liquid hydrogen, which has unmatched combustion performance and heat sink capability. Yet, liquid hydrogen's low density (1/11th that of hydrocarbons) and high handling cost for maintaining it at approximately 20 K leave plenty of room for alternative fuels for many applications that are incompatible with large (hydrogen) vehicles and a "hard cryogen" infrastructure. A good, long-term fuels program would need to be coordinated with vehicle development under the umbrella of a national program such as NAI, since the fuel selection is driven by vehicle mission and operational envelope and would benefit from coordination with the development of rocket propellants. The three main vehicle categories—expendables (missiles), accelerators (launch vehicles), and cruise—place quite different constraints on the fuel. The program discussed below is an attempt to present programs that support hypersonics in general rather than a specific application.

As a baseline, assume that both hydrogen and hydrocarbon flight test vehicles have accomplished successful flight demonstrations. The engineering required to convert the hydrocarbon flight test vehicle to a successful standoff weapon for attacking time-critical targets is considered to be outside the scope of this research proposal. The state of the art appears to be Mach 6-7 for this type of expendable application and current endothermic fuel technology. This is not to minimize the difficulty and expense of successfully completing these flight tests—far from it! It is just a recognition that these vehicles are too far along to benefit much from fuel research at this point. This plan also does not include any further liquid hydrogen development, such as densification, gelation, and the like.

For the proposed endothermic fuel research, efforts are described that culminate in demonstrations, but also included is the development of modeling and simulation rules and tools for the effective use of the improved technology. A notional timeline is shown in Figure E-1 and discussed below.

1. *Improved heat sink and coke reduction for conventional endothermic fuels.* This task would address the development of improved catalysts, additives, and coke-resistant designs for the n-octane/JP-7 kerosene class of hydrocarbons. (Rough order of magnitude [ROM] $500,000 per year for 5 years, with technologies demonstrated in long-duration panel tests in a radiant-heated AFRL facility)

2. *Robust supersonic ignition and combustion of hydrocarbon fuels.* This task would continue ongoing efforts to improve the combustion behavior of endothermic fuels, especially at the lower Mach numbers (including ignition). "Barbotage" fuel injection, fuel additives, ignition schemes, and high-performance piloting are options to improve endothermic fuel combustion and ignition performance. This is an expensive effort since it requires significant combustor testing under realistic conditions. (ROM $1 million per year for 7 years, culminating in a freejet test of an improved cold start/piloting system)

3. *Improved endothermic hydrocarbon physical properties.* It turns out that the physical properties of kerosene fuels are not well understood under high-temperature and high-pressure conditions. This work would be an extension of ongoing NIST work for NASA on RP-1. (ROM $200,000

[1] Tom Curran, former chief scientist of the (now) Propulsion Directorate of the Air Force Research Laboratory, at the 1988 NAS Woods Hole Conference "Fuels for Future High-Speed Flight Vehicles."

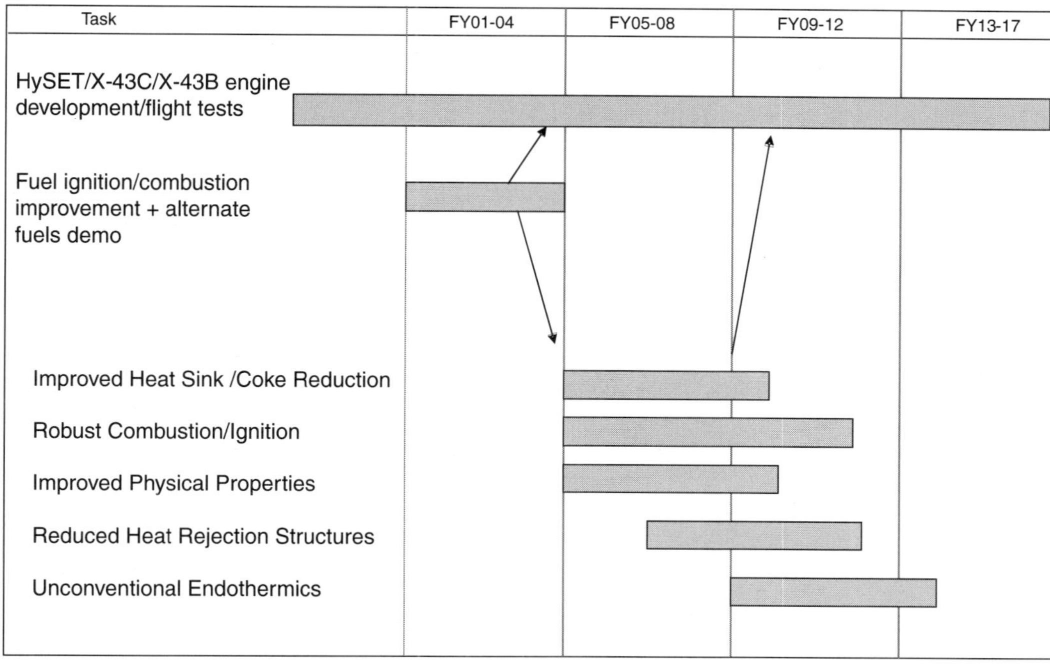

FIGURE E-1 Proposed endothermic fuel research timeline. SOURCE: Adapted from information provided by James T. Edwards, AFRL.

per year for 5 years, with the result being better physical property prediction tools for design and development)

4. *Development of reduced-heat-rejection structures.* To extend the Mach number capability of nonhydrogen-fueled hypersonic vehicles, one has to either improve the fuel heat sink capability or reduce the heat load from the vehicle into the fuel. The use of high-temperature ceramic engine structures appears to be capable of significantly reducing the required engine cooling (since the heat load into the fuel is proportional to the $T_{combustion}$ less the $T_{surface}$ driving force). Two key issues need to be addressed: (1) the structural manufacturability and durability of ceramic structures and (2) exposure of fuel to high-temperature surfaces and the impact on coking. (ROM $750,000 per year for 5 years, culminating in a direct-connect demonstration of a ceramic combustor)

5. *Evaluation of unconventional endothermic fuels.* This task would continue the development of steam reforming for increased heat sink (following up on current Russian work), as well as investigate other approaches to obtaining dramatically higher heat sinks than conventional endothermic fuels. For example, methanol reforming has been suggested as a promising alternative to steam reforming. This task would also look at more exotic alternatives, such as liquid metal decomposition and high-energy-density fuels. The fuel development would include both combustion and regenerative cooling research. (ROM $500,000 per year for 5 years)

REFERENCES

Published

Churchill, A.V., J.A. Hager, and A.E. Zengel. 1965. Fuels for Advanced Air-Breathing Weapon Systems. ASME Paper 65-0804. October. New York: American Society of Mechanical Engineers.

Huang, H., and D.R. Sobel. 2002. Endothermic Heat Sink of Jet Fuels. AFRL-PR-WP-TR-2002-2094. East Hartford, Conn.: United Technologies Research Center.

Huang, H., L. Spadaccini, and D. Sobel. 2002. Endothermic Heat-Sink of Jet Fuels for Scramjet Cooling. AIAA Paper 2002-3871. East Hartford, Conn.: United Technologies Research Center.

Lander, H., and A.C. Nixon. 1971. Endothermic fuels for hypersonic vehicles. Journal of Aircraft. 8(4):200-207.

Lehrach, R.P., L.J. Spadaccini, M.R. Glickstein, W.T. Fisher, and C.B. Graves. 1995. Applications Study of Endothermic Fuels. WL-TR-96-2122. East Hartford, Conn.: United Technologies Research Center.

Lipinski, J., R. Johnson, and C. White. 1992. Design, Fabrication, and Testing of a Lightweight Endothermic Methylcyclohexane Fuel Heat Exchanger/Reactor System. WL-TR-92-2076. East Hartford, Conn.: United Technologies Research Center.

Sobel, D.R., and L.J. Spadaccini. 1994. Development of Endothermic Potential of JP-8. WL-TR-94-2059. East Hartford, Conn.: United Technologies Research Center.

Spadaccini, L.J., M.B. Colket, P.J. Marteney, R. Roback, and M.R. Glickstein. 1993a. Endothermic Fuel/Catalyst Development and Evaluation Phase I. WRDC-TR-89-2141. East Hartford, Conn.: United Technologies Research Center.

Spadaccini, L.J., D.R. Sobel, M.B. Colket, P.J. Marteney, and M.R. Glickstein. 1993b. Endothermic Fuel/Catalyst Development and Evaluation Phases II, III, and IV. WL-TR-91-2126. East Hartford, Conn.: United Technologies Research Center.